SpringerBriefs in Applied Sciences and Technology

SpringerBriefs present concise summaries of cutting-edge research and practical applications across a wide spectrum of fields. Featuring compact volumes of 50 to 125 pages, the series covers a range of content from professional to academic.

Typical publications can be:

- A timely report of state-of-the art methods
- An introduction to or a manual for the application of mathematical or computer techniques
- A bridge between new research results, as published in journal articles
- A snapshot of a hot or emerging topic
- An in-depth case study
- A presentation of core concepts that students must understand in order to make independent contributions

SpringerBriefs are characterized by fast, global electronic dissemination, standard publishing contracts, standardized manuscript preparation and formatting guidelines, and expedited production schedules.

On the one hand, **SpringerBriefs in Applied Sciences and Technology** are devoted to the publication of fundamentals and applications within the different classical engineering disciplines as well as in interdisciplinary fields that recently emerged between these areas. On the other hand, as the boundary separating fundamental research and applied technology is more and more dissolving, this series is particularly open to trans-disciplinary topics between fundamental science and engineering.

Indexed by EI-Compendex, SCOPUS and Springerlink.

Raden Sjarief Widjaja · Hasanudin ·
Yuda Apri Hermawan · Azman Ismail ·
Fatin Nur Zulkipli · Andreas Öchsner
Editors

Marine Technology 4

Springer

Editors
Raden Sjarief Widjaja
Department of Naval Architecture
Institut Teknologi Sepuluh Nopember
Surabaya, Jawa Timur, Indonesia

Hasanudin
Department of Naval Architecture
Institut Teknologi Sepuluh Nopember
Surabaya, Jawa Timur, Indonesia

Yuda Apri Hermawan
Department of Naval Architecture
Institut Teknologi Sepuluh Nopember
Surabaya, Jawa Timur, Indonesia

Azman Ismail
Maritime Engineering Technology
and Centre for Women Advancement
and Leadership, Malaysian Institute
of Marine Engineering Technology
Universiti Kuala Lumpur
Lumut, Perak, Malaysia

Fatin Nur Zulkipli
College of Computing, Informatics
and Mathematics
Universiti Teknologi MARA
Machang, Kelantan, Malaysia

Andreas Öchsner
Faculty of Mechanical Engineering
Esslingen University of Applied Science
Esslingen am Neckar, Baden-Württemberg,
Germany

ISSN 2191-530X ISSN 2191-5318 (electronic)
SpringerBriefs in Applied Sciences and Technology
ISBN 978-3-032-08839-0 ISBN 978-3-032-08840-6 (eBook)
https://doi.org/10.1007/978-3-032-08840-6

This Springer imprint is published by the registered company Springer Nature Switzerland AG
The registered company address is: Gewerbestrasse 11, 6330 Cham, Switzerland

Preface

This book explores a variety of research topics in marine technology, delving into a wide array of crucial research topics within the maritime industry and its technology, encapsulating the latest research findings presented at The 8th International Conference on Marine Technology (SENTA 2023). It provides the exchange of recent knowledge, experiences, and innovations in the field of marine science and technology, encompassing naval architecture and technology; marine system and safety; ocean, coastal, and offshore engineering; shipping, port, and maritime logistics; underwater technology; and advanced technology in the maritime industry. This book covers various studies ranging from ship design and production techniques, comprising the ship design process, ship hydrodynamics, ship structure and advanced ship material, and ship management and production technology. The book also examines the invention in marine operation, marine machinery and maintenance systems, and marine safety, including digital technology in marine systems and safety. It also addresses the topics on coastal, ocean, and offshore technology, starting from coastal/offshore hydrodynamics, ocean energy, mooring line analysis, to offshore structure analysis. In addition, shipping, port, and marine logistic researches are also conveyed in this book, especially on shipping operation optimization, port development, green port, and smart shipping development. Moreover, underwater technology and advanced technology in the marine industry, i.e., computer vision technology for underwater vehicles, digital technology on ship design and production, and advanced computer technology in port and shipping development, have been discovered. Overall, this book offers a cradle for the exchange of ground-breaking ideas, fostering collaboration, and potentially setting the stage for significant developments in marine technology.

Surabaya, Indonesia	Raden Sjarief Widjaja
Surabaya, Indonesia	Hasanudin
Surabaya, Indonesia	Yuda Apri Hermawan
Lumut, Malaysia	Azman Ismail
Machang, Malaysia	Fatin Nur Zulkipli
Esslingen am Neckar, Germany	Andreas Öchsner

Contents

Innovative Technology for Automatic Solid Waste Mitigation at Estuaries to Prevent Marine Debris Entry into the Oceans

Nabiilah Aziizah Tjandra, Muhammad Hafiizh Imaaduddiin, and Mohamad Khoiri

Abstract Urban growth and industrial activities have led to a surge in environmental challenges due to improper waste disposal. In 2020, the influx of waste into oceans totaled 521,540 tons, predominantly comprising solid waste. This accumulation in estuaries and rivers has consequently contaminated marine ecosystems, imperiling biodiversity, and disrupting sectors like fishing and tourism. This study centers on a laboratory-scaled prototype of waste mitigation technology for estuarine implementation. The prototype aims to avert solid waste from entering oceans by utilizing a gate as a barrier and an automated filter for waste transportation and containment. The design factors waste type and daily weight, enabling calculations of gate and filter resistance and containment capacity. An automated system deploys limit switch sensors and infrared technology, managed by a microcontroller. Solar panels serve as an energy source, addressing power needs. Experimental outcomes reveal the filter's capacity to lift 5 kg of waste per cycle in 3 min. With a daily waste projection of 0.7 tons, potential waste reduction is up to 75%. Mitigating waste input anticipates minimizing negative impacts on biodiversity, coastal economies, and sustaining marine ecosystems for posterity.

Keywords Estuaries waste disposal · Solid waste mitigation · Automated technology

N. A. Tjandra (✉)
Department of Instrumentation Engineering, Faculty of Vocational Study, Institut Teknologi Sepuluh Nopember, Surabaya, Indonesia
e-mail: nabiilahtjandra@gmail.com

M. H. Imaaduddiin · M. Khoiri
Department of Civil Infrastructure Engineering, Faculty of Vocational Study, Institut Teknologi Sepuluh Nopember, Surabaya, Indonesia
e-mail: m_hafiizh@ce.its.ac.id

M. Khoiri
e-mail: mkhoiri@ce.its.ac.id

R. Sjarief Widjaja et al. (eds.), *Marine Technology 4*,
SpringerBriefs in Applied Sciences and Technology,
https://doi.org/10.1007/978-3-032-08840-6_1

1 Introduction

The ocean is crucial for Indonesians, providing essential food resources and livelihoods [1]. Safeguarding marine ecosystems, managing fisheries sustainably, controlling pollution, and conserving marine resources are vital measures [2]. Yet, a major challenge is the escalating issue of marine debris [3]. Originating from rivers, waterways, and coastal activities, it reached 521,540 tons in Indonesia in 2020, with over 50% being solid waste [4] [5]. A significant portion of this waste finds its origins in river mouths, exacerbating marine pollution [6]. The persistence of this problem can be attributed to inadequate waste management practices and suboptimal community behaviors [7]. This contamination poses severe threats to marine ecosystems, the transmission of diseases, and the overall environmental quality [8]. Pre-existing waste management technologies employed in water systems, such as the "trash boom" approach installed in rivers, serve to contain and facilitate the retrieval of debris [9]. However, this system has limited effectiveness in retaining debris for extended periods [10]. Previous research, such as [11] and [12], has implemented automatic innovation, demonstrating its effectiveness and necessity in addressing this issue. In this study, the author proposes a laboratory-scale prototype for innovative solid waste management. The technology, designed for estuaries, prevents solid waste from entering the ocean, controlled by an Arduino-based system with integrated solar panels for remote energy supply, ensuring reliability in estuarine locations [13].

2 Methodology

2.1 Literature Study

This research centers on two key questions: the necessity of automated technology in addressing waste management issues and the proposed device's capacity to tackle the problem [11]. Prior innovations, like the "trash boom", have shown limitations, such as the need for manual debris retrieval and potential leakage. The selection of components for the proposed device is guided by a comprehensive literature review, contextualizing the study within the broader research landscape.

2.2 Hardware Design

In developing the lab-scale waste retrieval technology, we adhered to outdoor material standards, utilizing corrosion-resistant galvalume for the mechanical framework. Filtering employs a 4 × 4 grid plastic-coated wire mesh, while the gate structure uses a 10 × 10 wire mesh (Fig. 1). Plastic bags and bottles clog estuaries, and we've

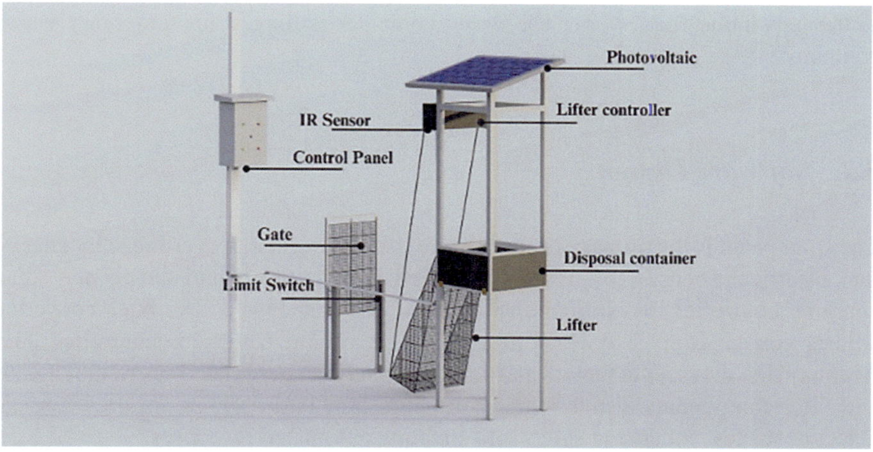

Fig. 1 Hardware design of the automatic waste lifter

Table 1 Waste mass

Waste	Mass (gram)
600 ml empty plastic bottle	30
600 ml filled plastic bottle	600

categorized the weights of empty and water-filled bottles in Table 1. Technology capacity is calculated based on the waste weight using the provided equation:

$$d = \frac{1}{2} \cdot a \cdot t^2 \tag{1}$$

where d is the lifting distance (m), a is the acceleration (m/s^2), and t is the time required to lift the load (s). And in this case, acceleration will be influenced by the weight of the load, the torque of the servo motor, and system characteristics. So, the acceleration is transformed to:

$$a = \frac{T}{m} \tag{2}$$

where T is the torque (N/m) and m is mass (kg), so the equation work in this research is:

$$d = \frac{1}{2} \cdot \frac{T}{m} \cdot t^2 \tag{3}$$

The technology comprises two components: the gate and the filter. The gate acts as a temporary barrier, directing waste to the filter, which conveys it to the waste container. The control panel is elevated above the device, with a solar panel attached to the galvalume framework. The waste container is hinged for temporary waste containment.

2.3 *Software Design*

The solar panel is the primary energy source, harvesting and converting solar energy into electrical energy [14]. Energy is stored in a battery, regulated by the solar charging controller for optimal charging, as shown in Fig. 2. The buck converter adjusts voltage from 5 to 12 V to match the Arduino Uno microcontroller. The Arduino Uno oversees actuators and sensors, ensuring coordinated operation of gate and filter components. Limit switches monitor boundaries, while infrared sensors detect obstacles, enhancing safety and enabling automated responses.

2.4 *System Testing*

Testing aims to validate seamless operation in both automatic and manual modes of a waste retrieval system, with success defined by flawless execution. Waste triggers the lifter via the upper limit switch, accommodating up to 5 kg per lift. The IR sensor monitors waste levels, closing the gate at a predetermined threshold. Simultaneously, the lower limit switch facilitates waste transfer into the container. Retrieval completes with the lifter descending, repeating the cycle for efficient waste management. The experiment also assesses lifting capacity with varying loads, correlating lifting time with load weight for adaptability and performance evaluation.

3 Results and Discussion

3.1 *Laboratory-Scale Prototype*

The laboratory-scale prototype is strategically positioned near a channel that collects rainwater runoff from a dam. This location facilitates rigorous testing with plastic bottle waste. Figure 3a provides a visual representation of the device's installation. The system includes a designated waste container, as depicted in Fig. 3c, offering temporary storage for accumulated waste. This feature enhances the overall efficiency and convenience of the waste management process, allowing for interim containment before manual retrieval and collection by human operators.

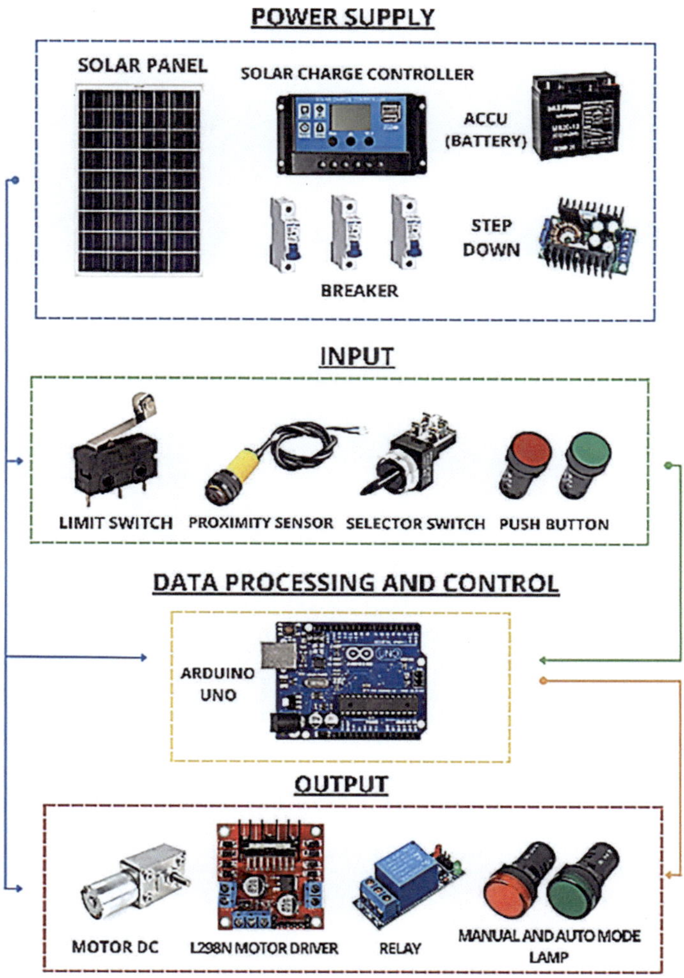

Fig. 2 Schematic diagram of the electrical components

3.2 System Running and Time Usage

The initial testing aimed to determine the filter's maximum lifting capacity by assessing its performance under various waste conditions, including empty, half-filled, and water-filled water bottles, as shown in Fig. 3b. The results in Table 2 revealed a maximum single lifting capacity of 5 kg, beyond which the lifter encountered difficulties, causing potential damage to the motor. Subsequent comprehensive testing assessed waste lifting capabilities and the impact of waste weight on lifting time. Using 600 ml plastic bottles in different conditions, the results in Table 3 show a minor effect on lifting duration, averaging approximately 3 min. These

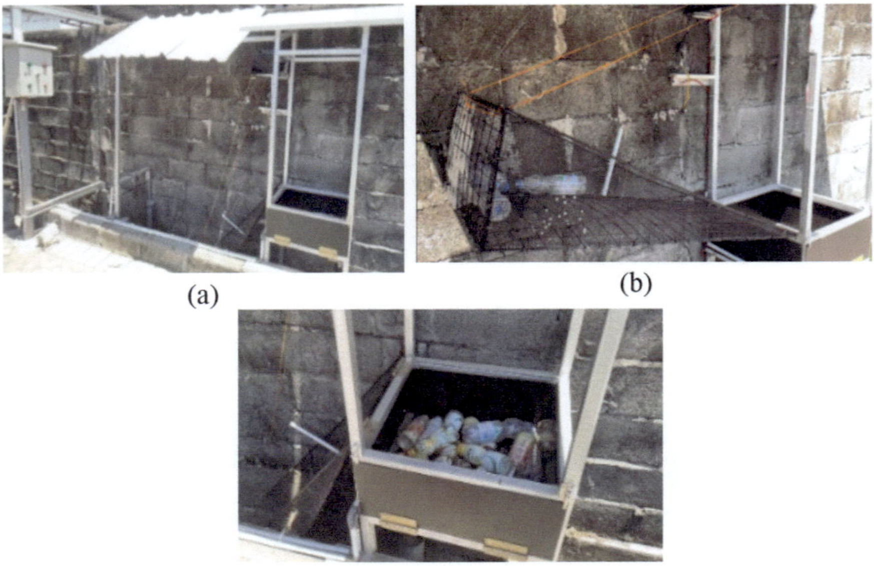

(a) (b) (c)

Fig. 3 **a** Laboratory-scale prototype, **b** lifter process, and **c** waste container

findings provide crucial insights into the system's operational limits and potential optimizations for efficient waste management.

Table 2 Test results of lifting capacity

Weight of lifted waste (kg)	Ability to lift
1	Yes
2	Yes
3	Yes
4	Yes
5	Yes
5,5	No
6	No

Table 3 Time usage

Waste mass (kg)	Lifter time (min.)	Total time (min.)
0.5	1	2
1.5	1.5	2
2.5	1.5	2
3	1.5	2
3.5	1.5	2,3
4	1.5	2.3
4.5	1.5	3
5	1.5	3

4 Conclusion

In summary, the laboratory-scale prototype for automated waste retrieval technology is a significant advancement in addressing outdoor waste management challenges. Material selection adheres to rigorous outdoor standards, ensuring durability. The pivotal roles of the IR sensor and limit switches enhance safety and precision. Practical experiments will validate the system's reliability in both modes, and testing will provide insights into lifting capacity and time for optimization. The prototype's installation in a real-world environment, near a water channel with rainwater runoff and plastic waste, allows for realistic testing, highlighting its potential contribution to efficient waste management and environmental sustainability.

Acknowledgements The authors gratefully acknowledge financial support from the Institut Teknologi Sepuluh Nopember for this work, under the project scheme of the Publication Writing and IPR Incentive Program (PPHKI).

References

1. D. Ferrol-Schulte, P. Gorris, W. Baitoningsih, D.S. Adhuri, S.C.A. Ferse, Coastal livelihood vulnerability to marine resource degradation: A review of the Indonesian national coastal and marine policy framework. Mar. Policy **52**, 163–171 (2015)
2. A. Gkanasos et al., Stopping macroplastic and microplastic pollution at source by installing novel technologies in river estuaries and waste water treatment plants: the CLAIM project. Front. Mar. Sci. **8**, 738876 (2021)
3. A.D. Sakti et al., Multi-scenario model of plastic waste accumulation potential in indonesia using integrated remote sensing, statistic and socio-demographic data. ISPRS Int J Geoinf **10**(7), 481 (2021)
4. M.R. Cordova, E. Riani, A. Shiomoto, Microplastics ingestion by blue panchax fish (Aplocheilus sp.) from Ciliwung Estuary, Jakarta, Indonesia. Mar. Pollut. Bull. **161**, 111763 (2020)
5. M.G. Gross, Effects of waste disposal operations in estuaries and the coastal ocean. Annu. Rev. Earth Planet. Sci. **6**(1), 127–143 (1978)

6. B. Pamungkas, R. Kurnia, E. Riani, Potential plastic waste input from Citarum River, Indonesia. Aquac., Aquar., Conserv. Legis. **14**(1), 103–110 (2021)
7. E.A. Wicaksono, A. Tahir, S. Werorilangi, Preliminary study on microplastic pollution in surface-water at Tallo and Jeneberang Estuary, Makassar, Indonesia. Aquac., Aquar., Conserv. Legis. **13**(2), 902–909 (2020)
8. H. Luo et al., Waste oyster shell as a kind of active filler to treat the combined wastewater at an estuary. J. Environ. Sci. **25**(10), 2047–2055 (2013)
9. M.A. Mallin, Impacts of industrial animal production on rivers and estuaries. Am. Sci. **88**(1), 26 (2000)
10. P.D. Queiroz, J.L.D. Dantas, F.S. Castro, G.M. Galvão, F.A. Mata, Development of numerical tool for loads and displacement prediction in a river debris removal solution, in *39th IAHR World Congress*, 2022
11. A. Gkanasos et al., Stopping macroplastic and microplastic pollution at source by installing novel technologies in river estuaries and waste water treatment plants: The claim project. Front Mar. Sci. **8** (2021). https://doi.org/10.3389/fmars.2021.738876
12. A. Winterstetter, M. Grodent, V. Kini, K. Ragaert, K.C. Vrancken, A review of technological solutions to prevent or reduce marine plastic litter in developing countries. Sustainability (Switzerland). **13**(9) (2021). https://doi.org/10.3390/su13094894
13. S. Kulaksız, M. Bau, Contrasting behaviour of anthropogenic gadolinium and natural rare earth elements in estuaries and the gadolinium input into the North Sea. Earth Planet. Sci. Lett. **260**(1–2), 361–371 (2007)
14. R.J. Cavagnaro et al., Powering the blue economy: Progress exploring marine renewable energy integration with ocean observations. Mar. Technol. Soc. J. **54**(6), 114–125 (2020)

Navigating Success: An In-Depth Study of Knowledge Sharing and Job Satisfaction Among Educators in Maritime Education and Training Institutions in Malaysia

Noorlee Boonadir, Rosnah Ishak, Roslin Ramli, and Noorazilah Ibrahim

Abstract This study aims to evaluate the levels of knowledge and job satisfaction among educators in maritime education and training institutions in Malaysia (MET). Validated questionnaires were employed to assess job satisfaction and knowledge management among MET educators, including various aspects of knowledge management such as creation, capture, sharing, distribution, acquisition, and application. Additionally, factors contributing to job satisfaction, including motivation and hygiene, were investigated to achieve the study's objectives. The results indicate that instructors in maritime institutions demonstrate a high level of proficiency in all components of knowledge management, with a high level of job satisfaction reported among the participants. The analysis reveals a moderate but significant relationship between knowledge management and job satisfaction, which is beneficial for the professional development of teaching staff in maritime institutions. Furthermore, the Ministry of Higher Education can use these findings to evaluate how educators in maritime institutions enhance knowledge management to meet job satisfaction standards and make informed decisions regarding structural improvements.

Keywords Knowledge management · Job satisfaction · MET

N. Boonadir (✉) · R. Ramli
Malaysian Institute of Marine Engineering Technology, Universiti Kuala Lumpur, Lumut, Perak, Malaysia
e-mail: noorlee@unikl.edu.my

R. Ramli
e-mail: roslin@unikl.edu.my

N. Boonadir · R. Ishak
Universiti Pendidikan Sultan Idris, Tanjong Malim, Perak, Malaysia
e-mail: rosnah.ishak@fpe.upsi.edu.my

N. Ibrahim
College of Computer, Informatics and Media, Universiti Teknologi MARA, Melaka, Malaysia
e-mail: azilahibrahim@uitm.edu.my

R. Sjarief Widjaja et al. (eds.), *Marine Technology 4*,
SpringerBriefs in Applied Sciences and Technology,
https://doi.org/10.1007/978-3-032-08840-6_2

1 Introduction

MET institutions have emerged as vital centers for maritime education and training, where a diverse range of academic disciplines, including business administration, information systems, management, and library sciences, are taught. Knowledge management (KM) systems, known for their effectiveness in corporate strategies, information technology, and human resource management across various sectors, are increasingly recognized as valuable tools for these institutions.

In MET institutions, it is imperative for the management, lecturers, academicians, and researchers to embrace efficient KM methods to enhance the quality of knowledge acquisition and dissemination. This process involves fostering a culture of open communication, encouraging the sharing and exchange of knowledge and information among individuals. Furthermore, MET institutions often seek consultation services to tailor KM strategies to align with their specific targets and objectives. The application of KM methodologies plays a critical role in helping organizations create, leverage, improve, and enhance their capabilities and knowledge resources [1].

The implementation of KM practices in these organizations supports knowledge sharing and enhances the decision-making process. The effectiveness of KM can be observed through an organization's approach to knowledge, encompassing its accumulation, utilization, sharing, and ownership [2]. This approach focuses on human knowledge management, organizational processes, and its impact on job satisfaction. Educational institutions, including MET institutions, serve as a balanced approach that intertwines human-oriented aspects with effective teaching and learning methods.

This section of the study delves into the knowledge management process and its relationship with job satisfaction, presenting the research model and hypotheses regarding the impact of KM processes on job satisfaction. The study employs the integrated KM cycle, Knowledge Management in theory and practice" of Kimiz Dalkir in [3] as illustrated in Fig. 1, and Herzberg's motivation hygiene theory, introduced in 1959, as depicted in Fig. 2.

2 Methodology

The main aim of this research is to explore the connection between job satisfaction and knowledge management in the context of educators working in maritime institutions in the Peninsular Malaysia. In addition, this study seeks to evaluate the extent to which each component of knowledge management is utilized. Furthermore, it intends to investigate how demographic factors such as gender, age, years of service, specific institutions, departments, and educational background impact both job satisfaction and knowledge management. This research employs a quantitative research approach to analyze the relationship between the independent variable, the components of

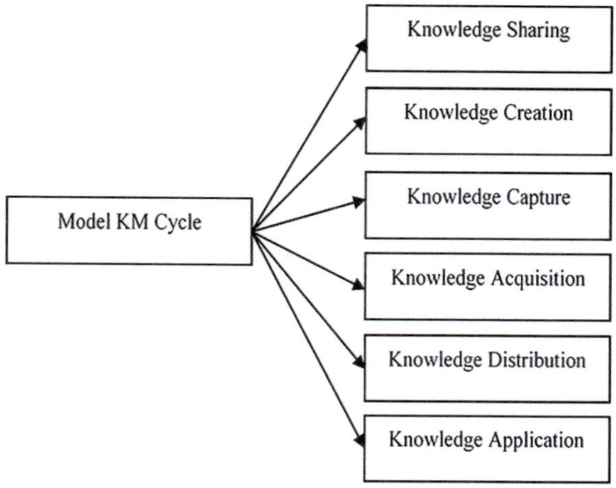

Fig. 1 The integrated KM cycle. Adapted from "knowledge management in theory and practice" of Kimiz Dalkir in [3]

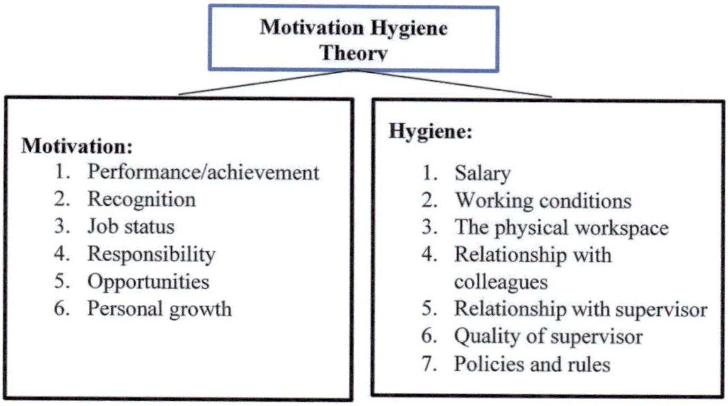

Fig. 2 Motivation hygiene theory. Adapted from "Motivation to Work." Of Herzberg [5]

knowledge management, and the dependent variable, job satisfaction, as indicated in Table 1.

The study's participants are drawn from academic professionals employed in maritime institutions across the Peninsular Malaysia, and a simple random sampling method is used to select the study's sample. Anticipated outcomes from this study are expected to provide valuable insights into the factors that influence job satisfaction and knowledge management among educators in the maritime sector. These insights

Table 1 Summary of variable, theory and model

Variable	Theory	Model
Knowledge management	The integrated KM cycle	Kimiz Dalkir, 2005
Job satisfaction	Motivation hygiene theory	Herzberg, 1959

have the potential to lead to improvements in working conditions [4] and the promotion of knowledge-sharing practices within maritime institutions in the Peninsular Malaysia.

2.1 Correlation Between Knowledge Management Factors and Enhanced Job Satisfaction Among MET Educators

The research conducted in maritimeinstitutions within the Peninsular Malaysia has yielded compelling evidence of a significant relationship between knowledge management factors and job satisfaction factors. In this study, the knowledge management factors under investigation included knowledge creation, knowledge capture, knowledge sharing, knowledge distribution, knowledge acquisition, and knowledge application. On the other hand, the job satisfaction factors were comprised of motivation and hygiene. The findings unequivocally point to a substantial and meaningful link between these knowledge management factors and job satisfaction, indicating that effective knowledge management practices have a positive influence on job satisfaction.

These outcomes align with prior research that has similarly underscored the interplay between knowledge management and job satisfaction. For instance, as proposed by [5], organizations that prioritize knowledge management practices are better positioned to comprehend and fulfill their employees' requirements, thus strengthening job satisfaction. Furthermore, [6] has suggested that when employees have access to essential resources and support, their levels of job satisfaction are significantly increased.

2.2 Significant Impact of Knowledge Management Practices and Systems

The significant discoveries arising from this study hold profound implications for the management of maritime institutions in the Peninsular Malaysia. By giving priority and investing in knowledge management practices and systems, organizations can elevate employee job satisfaction, motivation, and overall performance. Furthermore, the integration of job satisfaction factors, including motivation and hygiene, into knowledge management strategies can yield a more comprehensive and effective approach. Recognizing the main role of knowledge management practices in relation to job satisfaction equips educators with a deeper understanding of their responsibilities in efficiently managing knowledge resources and cultivating a positive work environment. This understanding serves to enhance their professionalism and allows them to make invaluable contributions to the overall success of the institution.

3 Results and Discussion

The study's results have unveiled that knowledge sharing represents the highest level of proficiency in knowledge management practices. This suggests that institutions are actively engaged in the dissemination and transfer of knowledge to enhance their operations and maintain competitiveness. As depicted in Table 2, these findings underscore the main role of knowledge management in the maritime industry, emphasizing that organizations giving it precedence are better poised for success in this sector.

These findings are in concordance with similar research conducted both in Malaysia and other nations, consistently showcasing a high degree of knowledge management practice across diverse sectors, including maritime and higher education institutions [3, 7, 8]. To stimulate the advancement of higher education institutions in Malaysia, it is imperative for the government to accord priority to knowledge management practices. This can be achieved by extending financial assistance and support to institutions, enabling them to implement effective strategies and practices for knowledge management. Given the significance of the maritime industry in terms of growth and sustainability, there should also be a notable emphasis on investment in research and development activities. This encompasses the exploration of novel technologies, processes, and best practices. Collaborations with industry and government entities, along with securing research funding from grants and various sources, can facilitate these endeavors.

To enhance education in the maritime sector, educators must be adequately equipped with ample resources and materials, including updated textbooks, online resources, and simulation software. Furthermore, financial support for research and development initiatives should be made available. Recognizing and rewarding outstanding educators through awards and accolades can serve as a motivating force for continuous improvements in teaching and research.

The observed moderate relationship between knowledge application and job satisfaction implies that employees who can effectively apply their knowledge in their job roles are more inclined to experience heightened levels of job satisfaction. This inclination may be attributed to their ability to witness how their knowledge and skills directly impact their work, resulting in a sense of accomplishment and fulfillment.

This robust association between knowledge application and job satisfaction finds support in prior research within the field of knowledge management. For example,

Table 2 Summary of research findings

No	Research question	Results/finding
1	What is the level of knowledge management element practices among academicians in MET institutions?	1. Knowledge sharing—high level 2. Knowledge creation—high level 3. Knowledge capture—high level 4. Knowledge acquisition—high level 5. Knowledge distribution—high level 6. Knowledge application—high level

a study by [9] discovered a positive and substantial correlation between knowledge application and job satisfaction among Iranian librarians. Similarly, another study by [9] revealed a positive connection between knowledge application and job satisfaction among nurses in the Philippines.

This discovery may be attributed to the understanding that mere access to information and knowledge may not be a direct driver of employee satisfaction. Instead, it is the capability to apply and share that knowledge that significantly influences job satisfaction. The research reinforces the idea that knowledge management practices should extend beyond merely storing and disseminating information. They should emphasize facilitating the use and application of knowledge to enhance both employee satisfaction and organizational performance.

Furthermore, the notion that recognizing the impact of one's knowledge and skills on their work can lead to a sense of accomplishment and fulfillment aligns with the concept of job crafting. Job crafting involves employees actively shaping their job responsibilities to better align with their strengths and interests. When employees can use their knowledge and skills to make meaningful contributions to their work, they are more likely to experience a sense of autonomy, competence, and relatedness, which are fundamental components of job satisfaction. This aligns with the findings demonstrating a significant relationship between knowledge application and job satisfaction in maritime institutions in the Peninsular Malaysia.

4 Conclusion

The importance of giving precedence to knowledge management practices in the evolution of higher education institutions in Malaysia, with specific emphasis on the maritime domain. It advocates for government involvement in terms of funding and support to aid institutions in the effective implementation of knowledge management strategies and practices. This support encompasses investments in research and development endeavors, including the creation of new technologies, processes, and best practices. Achieving this necessitates collaborations with industry and government bodies, as well as securing research funding from grants and various sources.

Moreover, the article underscores the necessity of ensuring maritime educators have access to ample resources and materials to fortify their teaching capabilities. Additionally, it underscores the importance of incentivizing excellence in teaching and research. By giving priority to knowledge management and research and development, the article contends that higher education institutions can make significant contributions to the growth and long-term sustainability of the maritime industry in Malaysia (Table 3) [10].

Table 3 Summary of research findings (null hypothesis)—Kruskal-Walis test and Man- Whitney U test

No		Hyphotesis null	Results/finding	Relationship
1	Ho1(a)	There is no positively relation between knowledge sharing with job satisfactions among educators in MET institutions	Rejected null hypothesis	Moderate

Acknowledgements The authors wish to extend their heartfelt gratitude and special appreciation to UniKL-Malaysian Institute of Marine Engineering Technology (UniKL-MIMET) for generously funding this research paper through the Research and Innovation section (R&I). They also wish to express their sincere thanks to their main supervisor at UPSI. Then, finally, the authors want to convey their warm regards to all individuals who have provided support and made valuable contributions to the successful completion of this paper.

References

1. B. Sripirabaa, S. Indumathi, Influence of organizational citizenship behavior on work performance: A study with specific reference to the bank employees in Coimbatore City. Manag. Today **7**(3), 145–157 (2017)
2. F. Torabi, J. El-Den, The impact of knowledge management on organizational productivity: A case study on Koosar Bank of Iran. Procedia Comput. Sci. **124**, 300–310 (2017). https://doi.org/10.1016/j.procs.2017.12.159
3. N.A. Hanafi, N.A. Ismail, Knowledge management practices in the shipping industry: An empirical study of Malaysian shipping companies. Int. J. Logist. Syst. Manag. **28**(1), 1–17 (2017)
4. Y.F. Chin, Malaysia: From hub to exporter of higher education and implications. Int. J. Bus. Soc. Sci. **10**(2), 48–54 (2019). https://doi.org/10.30845/ijbss.v10n2p6
5. Samwel, J, An assessment of the impact of performance management on employee and organization performance–evidence from selected private organizations in Tanzania. Int. J. Hum. Resour. Stud. 8(199) (2018). https://doi.org/10.5296/ijhrs.v8i3.13415
6. E. Demerouti, A.B. Bakker, F. Nachreiner, W.B. Schaufeli, The job demands-resources model of burnout. J. Appl. Psychol. **86**(3), 499–512 (2001)
7. R.M. Yusoff, N.I. Jaafar, H.A. Hashim, Knowledge management practices in Malaysian public universities. Malays. J. Libr. Inf. Sci. **22**(1), 69–86 (2017)
8. A. Ali, S. Ismail, Knowledge management practices in the Malaysian ports industry. J. Asia Pac. Econ. **24**(1), 69–86 (2019). https://doi.org/10.1080/13547860.2018.1477386
9. Gholami, R., Hosseini, S.M., Mohammadi, H, Relationship between knowledge management processes and job satisfaction: Case study of Iranian librarians. Libr. Philos. Pract., pp 1–17 (2015)
10. M.R. Mariano, G.G. Peralta, R.G. Trigo, Knowledge management, organizational support and job satisfaction among nurses in the Philippines. J. Nurs. Manag. **27**(1), 50–57 (2019)

Malcolm in the Middle: Determining the Alpha Individual in the Middle of Blacktip Reef Shark Schooling in Mitita Island, Morotai Waters

Aditya Nurfauzan, Darmawan Ahmad Mukharror, and Sunarto

Abstract The blacktip reef shark (*Carcharhinus melanopterus*) behavior has been researched since the 1950s. Being considered as one of the most abundant shark species in Indonesia, the blacktip reef shark plays a significant role in maintaining the health of coral reef ecosystems. With its status as the fourth largest country in the shark tourism economy and the largest shark catcher in the world, Indonesia tries to increase the awareness of conservation of sharks, and at the same time reduce shark catches. As one of the important shark diving locations, Morotai Waters offers a 100% sighting probability of blacktip reef sharks. Therefore, increasing the understanding of blacktip reef sharks' behavior and social structure in Mitita Island of Morotai Waters, is important to both conservation and tourism safety efforts. Determining the alpha individual is the key to mapping the social structure of blacktip reef shark communities. This study identified BRS-017 as the potential alpha individual in the shark habitat of Mitita Island. However, complex relations between school members have not been graphed. Hence, further research is required to enable a complete social structure of this shark schooling.

Keywords Blacktip reef shark · Mitita Island · Alpha individual · Social structure · Shark behavior

A. Nurfauzan (✉) · Sunarto
Marine Science, Faculty of Fisheries and Marine Science, Universitas Padjadjaran, Bandung 45363, Indonesia
e-mail: adityanf44@gmail.com

Sunarto
e-mail: sunarto@unpad.ac.id

D. A. Mukharror
Chemical Engineering Department, Faculty of Engineering, University of Indonesia, Depok 16424, Indonesia
e-mail: processsafetyindonesia@gmail.com

R. Sjarief Widjaja et al. (eds.), *Marine Technology 4*,
SpringerBriefs in Applied Sciences and Technology,
https://doi.org/10.1007/978-3-032-08840-6_3

1 Introduction

Sharks are philopatric animals, or these animals will return to the area where they came from, and when they are in their teens, they will remain in their forbidden area [1]. Blacktip reef sharks (*Carcharinus melanopterus*) are often found in coastal coral reef areas. This type of shark is found very abundantly on islands and coral atolls [2]. Spatial separation of the sexes is a widespread phenomenon in sharks [3] and can be categorized into two types: habitat segregation and social segregation. Habitat segregation occurs when the sexes differ in their habitat use, and social segregation is the tendency of a species to form single-sex groups [4].

Changes in the dynamics of social structures in living things will always occur across space and time so that they will make changes to the composition of groups and group experiences [5]. In the case of lemon sharks, the group leader will always have a larger body posture than individuals in the other group [1]. For sharks, seen from the nature of agonists, the nature of agonism is a competitive behavior and is important for understanding social dynamics in animals of many taxa. Every living creature has its agonist properties, and there have been several studies on the nature of agonists. Research on agonist properties was previously conducted by Martin in 2007 [6].

This research aims to rank shark dominance according to decision-making by looking at how many blacktip reef sharks treat the more dominant sharks during schooling, and this information will later be useful for looking at the form of the social structure of the blacktip reef shark on Mitita Island, Morotai Islands.

2 Methodology

2.1 Area Studies

The data collection location is on Mitita island, Morotai islands, with coordinates N 01° 58′ 13.4” E 128° 14′ 03.6”. This consideration refers to previous research regarding the blacktip reef shark, which was seen to be schooling in the area [6–8].

2.2 Data Collection Plan

The method used in this research is dive-operated video with a scheme of three divers who will collect data. One of the divers is a professional and understands how to feed sharks to lure them into the data collection area. The camera used here is a GoPro 10 Hero.

2.3 Individual Markers

To determine natural markers, this study uses references from previous research in 2019 conducted in the blacktip reef shark (*C. melanopterus*) region, with the results of the study showing that there are different patterns or patterns on the dorsal and caudal fins [8].

2.4 Data Analysis

Dominant Hierarchy Analysis

David score is a data analysis technique for calculating the probability of ranking dominance over the entirety of all interacting individuals [10]. DS has been used widely in social mammals, especially primates [9, 11]. The calculation of the David score is based on the number of wins and losses of an observed interaction. First, the proportion of wins in the interactions that occur is calculated. The winning proportion of an individual marked as *i* in its interaction with another individual *j* (P_{ij}) is the number of times individual *i* defeats *j* (S_{ij}) divided by the total number of interactions between *i* and *j* (n_{ij}) resulting in the following formula: (P_{ij}) is the number of times individual *i* beats *j* (S_{ij}) divided by the total number of interactions between *i* and *j* (n_{ij}) resulting in the following formula:

$$P_{\mathrm{ij}} = \frac{s_{\mathrm{ij}}}{n_{\mathrm{ij}}} \tag{1}$$

The proportion of individuals losing *i* in their interaction with *j* (P_{ij}) is equal to 1–*P*ij. If $\mathrm{n}_{ij} = 0$ then $P_{\mathrm{ij}} = 0$ and $P_{\mathrm{ij}} = 0$ (David's formula) DS for each individual *i*, a group is calculated by the formula:

$$\mathrm{DS} = w + w_2 - I - I_2 \tag{2}$$

To obtain a measure of steepness that varies between 0 and 1, DS needs to be converted to normalized DS (NormDS) as follows.

$$\mathrm{NormDS} = \frac{\{\mathrm{DS} + N(N-1)\}/2}{N} \tag{3}$$

The value of *N* is the maximum potential or number of individuals recorded. The wins at each number two are distributed completely unidirectionally (*i* for each number two (*i,j*) either *i* or *j* has won all dominance encounters between *i* and *j*). Then the wins and losses are presented in a matrix containing individuals that have been sorted from the highest rank [11].

3 Results and Discussion

3.1 Natural Markers

In 2019, 21 individuals were recorded as being present on one side, then the latest discovery in 2023 recorded that there were 18 individuals with a total of seven missing individuals and four new individuals with the following marking pattern: There are additional patterns and sizes for all individuals with the largest size being 157 cm (BRS—017) and the smallest size being 100 cm (BRS—022).

3.2 Social Structure Analysis

Visualization results show that the largest node size is in individual BRS—017 with a score of 12,039. The number of wins is calculated by the number of agonist traits during video recording that are captured, namely, train with a total of 48 train events. Of all the individuals present, the male with the code BRS—022, with a total train of four occurrences, most frequently follows this individual, and for the female, it is BRS—008. With a total of five train occurrences. However, among the 18 individuals, it was not apparent that individual BRS—025 followed BRS—017. The results found suggest that size has a considerable influence on hierarchical dominance, which can be seen from its agonistic nature.

4 Conclusion

The dominant individual according to individual ranking on blacktip reef shark habitat in Mitita waters is BRS—017 with DS norms 12,039, and second place is BRS—009 with DS norms 9693. The score has previously been calculated between the number of trains and giveaways. The largest node is shown by BRS—017, followed by BRS—009 and BRS—018 (Figs. 1 and 2, Table 1).

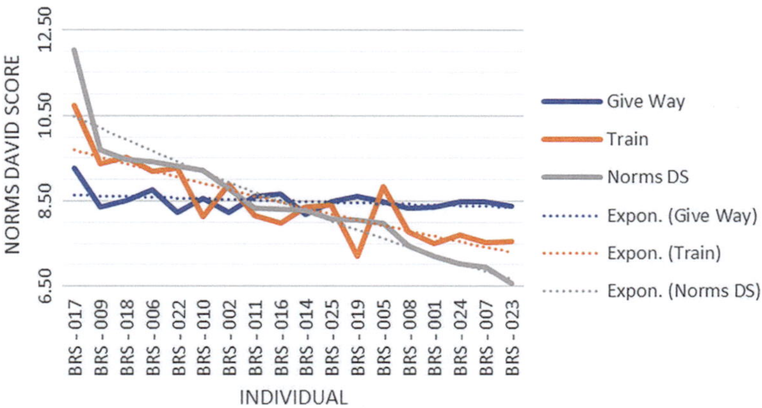

Fig. 1 Steepness according to norms DS

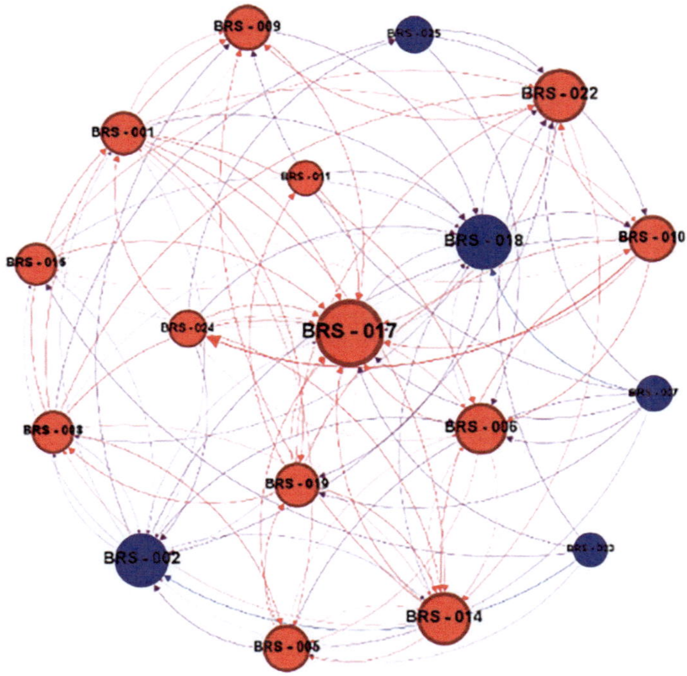

Fig. 2 Hierarchical dominance structure in alpha sharks with the largest nodes having the most potential to win

Table 1 Sizes and norms DS on blacktip found

Individual	Gender	DS	Size	Norms DS
BRS–001	F	− 23.57	145	7.19
BRS–002	M	5.26	137	8.79
BRS–005	F	− 9.45	134	7.98
BRS–006	F	16.65	152	9.42
BRS–007	M	− 28.01	143	6.94
BRS–008	F	− 19.06	154	7.44
BRS–009	F	21.48	130	9.69
BRS–010	F	13.14	138	9.23
BRS–011	F	− 2.93	142	8.34
BRS–014	F	− 3.64	137	8.30
BRS–016	F	− 3.59	146	8.30
BRS–017	F	63.71	156	12.04
BRS–018	M	17.34	138	9.46
BRS–019	F	− 7.85	140	8.06
BRS–022	M	14.74	100	9.32
BRS–023	M	− 34.99	115	6.56
BRS–024	F	− 26.50	102	7.03
BRS–025	M	− 7.37	101	8.09

Acknowledgements This project is funded by Shark Diving Indonesia Morotai, under the Department of Research. Therefore, the writers thank for the assistance given by all Shark Diving Indonesia Morotai's members during the research.

References

1. J. Mourier, S. Planes, Direct genetic evidence for reproductive philopatry and associated fine-scale migrations in female blacktip reef sharks (Carcharhinus melanopterus) in French Polynesia. J. Mol. Ecol. **22**, 201–214 (2013). https://doi.org/10.1111/mec.12103
2. Y.P. Papastamatiou, J.E. Caselle, A.M. Friedlander, C.G. Lowe, Distribution, size frequency, and sex ratios of blacktip reef sharks Carcharhinus melanopterus at Palmyra Atoll: A predator-dominated ecosystem. J. Fish Biol. **75**, 647–654 (2009). https://doi.org/10.1111/j.1095-8649.2009.02329.x
3. V.J. Wearmouth, D.W.Sims, Chapter 2 sexual segregation in marine fish, reptiles, birds and mammals. Behaviour Patterns, Mechanisms and Conservation Implications. Ad. In Marine. Bio. **54**, 107–170 (2008). https://doi.org/10.1016/S0065-2881(08)00002-3
4. J. Mourier, S.C. Mills, S. Planes, Population structure, spatial distribution and life-history traits of blacktip reef sharks Carcharhinus melanopterus. J. Fish Biol. **82**, 979–993 (2013). https://doi.org/10.1111/jfb.12039

5. A. Sih, S.F. Hanser, K.A. McHugh, Social network theory: New insights and issues for behavioral ecologists. Behav. Ecol. Sociobiol. **63**, 975–988 (2009). https://doi.org/10.1007/s.00265-009-0725-6
6. D.A. Mukharror, R. Fadli, F. Ghiffari, The COVID-19 pandemic and the impact on the absence of diver-shark interaction on blacktip reef shark's agonistic behavior: Morotai case, in *IOP Conference Series: Earth and Environmental Science, IOP Publishing Ltd 14 Nov 2020*
7. D.A. Mukharror, I.T. Baiti, M. Ichsan, N. Pridina, S. Triutami, Correlation between existence of reef sharks with abundance of reef fishes in south waters of Morotai Island (North Moluccas). in *IOP Conference Series: Earth and Environmental Science. Institute of Physics Publishing 22—25 Aug 2017*
8. D.A. Mukharror, D. Susiloningtyas, T. Handayani, N. Pridina, Blacktip reefshark (Carcharhinus melanopterus) individual's identification in Morotai waters using its fin's natural markings. in *AIP Conference Proceedings. American Institute of Physics Inc. 27 Dec 2019*
9. M.P. Gammell, H. De Vries, D.J. Jennings, C.M. Carlin, T.J. Hayden, David's score: A more appropriate dominance ranking method than Clutton-Brock et al.'s index. Anim. Behav. **66** (2003). https://doi.org/10.1006/anbe.2003.2226
10. H. De Vries, J.M.G. Stevens, H. Vervaecke, Measuring and testing the steepness of dominance hierarchies. Anim. Behav. **71**, 585–592 (2006). https://doi.org/10.1016/j.anbehav.2005.05.015
11. K.N. Balasubramaniam, C.M. Berman, A. De Marco, K. Dittmar, B. Majolo, H. Ogawa, B. Thierry, H. De Vries, Consistency of dominance rank order: A comparison of David's scores with I&SI and Bayesian methods in macaques. Am. J. Primatol. **75**, 959–971 (2013). https://doi.org/10.1002/ajp.22160

Web-Based Warship Reliability Assessment System

David Artanto, Trika Pitana, Udisubakti Ciptomulyono, and Muhammad Azka Bintang Pramudya

Abstract The Republic of Indonesia is an archipelagic nation with a territorial water area larger than its land area. This condition results in high demand for a fleet of warships that possess a high level of reliability to safeguard its territory. To meet the requirement for reliability in the fleet of warships, periodic monitoring is necessary to ascertain the reliability status of a warship as it carries out its duties as a crucial component in safeguarding the waters of the Republic of Indonesia. Additionally, regular maintenance in a timely manner and in accordance within the ship's needs is essential. The significance of warships in safeguarding Indonesia's maritime territory cannot be disregarded. Hence, planned monitoring and regular maintenance play a pivotal role in maintaining ship performance. Unplanned maintenance and monitoring can disrupt the operational capabilities of warships, jeopardizing the integrity of maritime security. In this disclosure, the researcher has developed a web-based information system to manage the reliability of warships. This system will categorize ship systems and subsystem in accordance with the applicable military standards. With the implementation of this information system, it is expected that the monitoring and maintenance of warships can be conducted more efficiently.

Keywords Warships · Reliability · Web-based information system

D. Artanto (✉) · T. Pitana · M. A. B. Pramudya
Department of Marine Engineering, Institut Teknologi Sepuluh Nopember, Surabaya, Indonesia
e-mail: davidartanto@gmail.com

T. Pitana
e-mail: trika@its.ac.id

M. A. B. Pramudya
e-mail: abinpramdya2@gmail.com

U. Ciptomulyono
Industrial Engineering Department, Institut Teknologi Sepuluh Nopember, Surabaya, Indonesia
e-mail: udisubakti@gmail.com

R. Sjarief Widjaja et al. (eds.), *Marine Technology 4*,
SpringerBriefs in Applied Sciences and Technology,
https://doi.org/10.1007/978-3-032-08840-6_4

1 Introduction

The condition of equipment within an operational system is the most important factor in determining the level of material readiness. Readiness comprises both the extent of apparatus damage and the rate at which it can be repaired when damage occurs. In addition, the equipment readiness model can be characterized as the consequence of the level of damage, maintenance capability, and supply support, which can be represented by the term equipment readiness [1].

From the perspective of a warship's system/subsystem component, one system is interconnected with another to form a larger and more complex system, including thrusters, electrical systems, navigation systems, communications, and combat management systems. Analysis involving these extremely complex components is a separate problem factor that necessitates the use of an appropriate analysis method so that the resultant analysis results can be used as preliminary information in the development of an operation plan, budget, and maintenance plan. This method of analysis must be objective, based on verifiable facts and data. In addition, an analysis process must be systematic, that is, its phases must be logical and organized. In the meantime, the analysis's results are informative, providing information that can be used to solve a problem or answer a query. The analysis yields valid results, which are consistent with extant facts and data and are not based on assumptions or conjecture. In addition, the results of the analysis are pertinent because they relate to the objective and context of the analysis procedure assessment.

When in the area of operations, it is difficult for warships to provide information about their combat capability due to the nature of their mission. Due to the delay in information, this system's primary obstacles are accuracy and the continuity of the conventional reporting system. A web-based information system is a technology that uses the internet to provide and manage diverse forms of information. Web-based information systems have several benefits, including increasing the accessibility of information for users in different locations and at different times, ensuring the availability and quality of relevant, timely, and accurate information, increasing productivity and operational efficiency for users and system developers, and representing a long-term investment in system development and maintenance.

A website's primary objective was to be a shared information space through which people could communicate using devices. It also provided a means for individuals to store and organize their own information in a way that would make it accessible to both them and others [2]. A web-based information system is one that uses web technologies on the internet to provide services and data to users. This technology, which is a software system, uses the hypertext principle to manage and retrieve information. A web browser is used as the front end and all the databases serve as the back end of a web-based database system, which is a combination of specific functionality-oriented elements of one or more web applications [3].

The purpose of this study is to conduct research on a system that can increase the Navy's efficiency in the warship assessment process in accordance with applicable standards, which refer to the TNI AL materiel assessment technical guidelines. This research proposes using a website-based platform that not only has been modified to meet the assessment requirements pertinent to the Navy environment, but also secured prior to earlier studies. This method has practical benefits and can provide information in real time.

2 Methodology

2.1 Ship Reliability Analysis

A warship is fundamentally segregated into two primary components. Firstly, the platform encompasses the ship's main structure, propulsion system, safety features, and support equipment. Secondly, there is the "sewaco", a shortage of sensors, weapons, and command employed by TNI AL to encompass the warship's equipment, such as armament, sensor arrays, as well as navigational and communication equipment. At first each piece of equipment on-board the vessel is initially evaluated by a specialist or related field crew member by using the assessed equipment historical data of maintenance and runtime and its manual book, if the equipment doesn't have a historical track record for its assessment, the crew may use the TNI AL guidelines book, which contains grading rules as shown on Table 1. The grades of each piece of equipment are then sent to the relevant higher-ups once it has been evaluated. The authorized higher-up will then access the web to input the grades for each assessed equipment into its respective subsystem where the equipment is installed. Once the website has been updated with the equipment's grade data, the formula (1) given will be applied to calculate the ship reliability.

$$R(e) = \mathrm{EG} \times W \tag{1}$$

$R(e)$ equipment reliability
EG equipment reliability grade
W equipment weight within its subsystem

At first, the formula above is applied to the grading of the assessed equipment; each equipment has its unique equipment weight for its subsystem. This weight represents the equipment's significance to its each respective subsystem reliability. The formulas used to determine the subsystem reliability score are as follows:

$$R(s) = \Sigma\left(R_{(e)} \times W\right) \tag{2}$$

$R(s)$ total subsystem reliability
$R(e)$ total equipment reliability
W system weight within the warship

Table 1 TNI AL equipment grading rules

Equipment grading rules		
Titles	Grade range (%)	Description
Very good	90–100	New equipment that's recently installed
Good	80–89	New equipment with warranty or recently fixed
Fair	65–79	Equipment that has been installed for a while and functional
Lightly damaged	50–64	Equipment with primarily functional components
Severely damaged	35–49	Equipment with partially working function
Heavily damaged	0–34	Equipment that is either fully damaged or barely functional

After getting the total equipment reliability, the formula above will be used to get the reliability of the subsystem within the warship, whereas the total equipment reliability is summed and then multiplied by each system's weight within the warship. This weight represents the significance of the subsystem for the warship's reliability. Finally, the following formula will be used to calculate the warship reliability scores.

$$R_{(t)} = \Sigma R_{(s)} \tag{3}$$

$R(t)$ total warship reliability
$R(s)$ total subsystem reliability

In this formula, the warship reliability is calculated by summing all of the subsystems within the vessel. After this calculation, the total reliability of the warship is returned by the website algorithm and will be available for any authorized user, whether within the vessel or not. The current ship reliability score can then be assessed by the appropriate stakeholders to determine the vessel's next operation and maintenance.

2.2 *Web User Authorization*

Within the website, various authorized users have distinct permissions and access levels. These include platform managers, sewaco managers, vessel commanders, and onshore administrators, with specific details outlined in Table 2. Each user's access to information and abilities is intentionally limited for two primary reasons: security and optimization.

In terms of security, this approach ensures that in the event of a breach, the intruder can only access a restricted set of information, making it easier to detect unauthorized access. Furthermore, from an optimization standpoint, specialized crew members proficient in specific roles can focus solely on their designated areas of expertise, minimizing the risk of errors and maximizing efficiency. Additionally, individualized

Table 2 Website account and authority

Website authority		
Account name	Account authority	Account description
Platform officer	View and edit the platform subsystem within the vessel it assigned to	The account for the platform officer of a vessel
Sewaco officer	View and edit the sewaco subsystem within the vessel it assigned to	The account for the sewaco officer of the vessel
Vessel commander	View every subsystem on the vessel it assigned to	The account of the vessel commander
Onshore administrator	View every data available of every vessel available	The account of an onshore admin mainly the headquarters

access for each account serves as a preventive measure against potential discrepancies or unauthorized actions. This means that any attempt to manipulate data or settings would either be promptly detected or, in the absence of proper authorization, simply impossible to carry out [4].

This solution, when applied within the website, means, a particular model of role-based access control is implied. Thus, according to Osborn, is has a great success when implemented on the commercial control systems that implement roles for security administration, each with its own authorization [5].

3 Results and Discussion

In this section, the researcher will illustrate the practical application of the website through various simulated scenarios. Specifically, the emphasis will be on modifying the assessed grades for the platform section equipment of the ship. This demonstration serves a dual purpose: firstly, to validate the previously presented formulas, and secondly, to exemplify how alterations in specific components can significantly influence the overall reliability scores of a warship, reflecting their respective importance to the vessel's performance. In this demonstration, the primary focus will be on manipulating the assessed grades of the following key equipment components, including two main diesel engines and five diesel generators.

The remaining equipment will maintain a standardized value of 80%, placing them within the "good" category to simulate the performance of a properly functioning vessel. In the first scenario, both primary components will be assigned a value of 80, signifying that both pieces of equipment are fully functional. When coupled with the remaining equipment, each is also rated at 80%. By using this scenario, the website calculates an overall high reliability score for the ship, as depicted in Fig. 1.

Fig. 1 Simulation 1 results

In the second scenario, we will replicate a simulated failure of the main diesel engine of the vessel. For the purposes of this demonstration, we will assign a low assessment grade to the diesel engine, specifically setting it at 40%, categorizing it as "severely damaged". It is important to note that all other equipment, including the diesel generators, will retain their high-grade rating of 80%. By using this scenario, the website calculates a reduced score of reliability score for the vessel, especially for the platform section, as depicted in Fig. 2.

In the third scenario, we will mimic a simulated failure of the diesel generator on-board the vessel. As in the second scenario, a low assessment grade of 40% will be assigned to the majority of the diesel generators, categorizing them as "severely damaged". However, it is worth noting that the rest of the equipment, including the diesel engine, which had previously received a low assessment grade in the last scenario, will now revert to their original rating of 80%. By using this scenario, the website calculated an even more reduced score of reliability for the vessel, especially for the platform section, as depicted in Fig. 3.

In the fourth scenario, we will simulate a dual failure involving both the vessel's main engine and the onboard diesel generator. Both pieces of equipment will receive low assessment grades of 40%, placing them in the "severely damaged" category.

Fig. 2 Simulation 2 results

Fig. 3 Simulation 3 results

Fig. 4 Simulation 4 results

As in the previous demonstration, the remaining equipment will maintain its high assessment grades of 80%. By using this scenario, the website calculated a severely low grade of reliability for the vessel, as depicted in Fig. 4.

4 Conclusion

In this paper, the researcher introduces a novel concept: a web-based information system tailored for the Indonesian Navy. Drawing from the established guidelines provided by the Indonesian Navy, this website incorporates the existing rules and assessment procedures, automating them with the aim of achieving real-time reporting and enhancing user-friendliness for all involved parties. Through the implementation of the web-based information system outlined in this paper, it is anticipated that the Indonesian Navy will find it more streamlined to track the maintenance of each vessel at their disposal. This advancement is expected to significantly bolster the safety and security of the nation of Indonesia in the process.

The study presented in this paper requires further extensive research and testing before it can be implemented in real-life scenarios and integrated into the fleet of warships. One critical area that demands heightened attention is the security of the website, considering it contains crucial information pertaining to every warship. This aspect must undergo rigorous enhancements to ensure its robustness and reliability, should the day come when this web-based information system is formally deployed in real-world applications for the Indonesian navy. Further development of this research is necessary to assess the combat capability of warships based on reliability. This pertains to ensuring the security of data stored on web servers, as well as preventing acts of data theft and sabotage. The objective is to ensure the data security of warship combat capability. There are several aspects that require inclusion in this context, namely user access and associated restrictions, alongside the provision of a network in cases where the vessel in operation is situated beyond the reach of the internet.

References

1. L.J. Junor, J.S. Oi, A new approach to modeling ship readiness. Cent. Nav. Anal. (1996)
2. R.T. Fielding, R.N. Taylor, Principled design of the modern web architecture. ACM Trans. Internet Technol. **2**(2), 115–150 (2002). https://doi.org/10.1145/514183.514185
3. Y.Y. Aung, Implementing a web-based information system. Co-operative University Sagaing (2018)
4. M. Bugliesi, S. Calzavara, R. Focardi, Formal methods for web security. J. Log.Al Algebr. Methods Program. **87**, 110–126 (2017). https://doi.org/10.1016/j.jlamp.2016.08.006
5. S. Osborn, R. Sandhu, Q, Munawer, Configuring role-based access control to enforce mandatory and discretionary access control policies. in *The University of Western Ontario and George Mason University*. ACM Trans. Inf. Syst. Secur. (TISSEC)ACM Trans. Inf. Syst. Secur. (TISSEC)ACM Trans. Inf. Syst. Secur. (TISSEC). **3**, 85–106 (2000). https://doi.org/10.1145/354876.354878

Analysis of Potential Value Enhancement of Domestic Component Levels for the Design and Development of a Remotely Operated Underwater Vehicle Power Supply

Andhika Asmara, Sjarief Widjaja, and Wasis Dwi Aryawan

Abstract Indonesia's domestic potential related to maritime, exploration and management of underwater areas is important. The most suitable vehicle infrastructure to achieve this goal is the development of underwater vehicles such as remotely operated vehicles (ROV). The level of domestic content (TKDN) is important in the industry so that it can run independently, so it is necessary to identify the potential for domestic supply and manufacture of these ROV components, such as ROV power supply. In identifying and calculating components, a gap analysis will be carried out related to existing components after conducting a market survey and analysis of all component values to achieve minimum domestic content of 40%. In the TKDN analysis method, firstly, we need to identify the components for ROV electrical and power supply. Secondly, a market survey and market analysis were carried out of available components on the market. Thirdly, analysis of the TKDN of each component. Fourth, optimization is carried out to determine the best component TKDN value and price assessment. Finally, the calculation of the overhead cost of ROV construction is carried out. The output of the analysis will provide strategies related to efforts to increase TKDN values in domestic ROV procurement.

Keywords Gap analysis · Power supply · ROV · TKDN

A. Asmara (✉) · S. Widjaja · W. D. Aryawan
Department of Naval Architecture, Institut Teknologi Sepuluh Nopember, Surabaya, Indonesia
e-mail: andhika2000asmara@gmail.com

S. Widjaja
e-mail: sjarief@na.its.ac.id

W. D. Aryawan
e-mail: wasis@na.its.ac.id

R. Sjarief Widjaja et al. (eds.), *Marine Technology 4*,
SpringerBriefs in Applied Sciences and Technology,
https://doi.org/10.1007/978-3-032-08840-6_5

1 Introduction

Indonesia is the largest archipelagic country with a total of 17,499 islands. The total area of Indonesia is 1905 million km^2, of which two-thirds of the territory is ocean. Therefore, underwater exploration and exploitation activities are very important to support the country's economy and infrastructure. Most of the Indonesian archipelago has large reserves of petroleum and natural gas. Based on data from the 2018–2027 Indonesian Natural Gas Balance by the Ministry of Energy and Mineral Resources of the Republic of Indonesia, it is projected that Indonesia can produce around 7000 MMSCFD (million standard cubic feet per day), whereas existing production sources only produce around 3000 MMSCFD. So, there is a need to increase production to meet Indonesia's domestic needs [1]. Remotely operated vehicles (ROVs) have been a frequent monitoring tool for subsea operations in the offshore oil and gas industry for more than the last 30 years [2]. The ROV is connected to a platform that functions to give commands, which is connected by a cable [3]. Seeing the potential that has been described, especially in Indonesia, is one of the aspects that encourages the provision and development of ROV units in the country. In procuring components related to the development and operation of domestic ROVs there are challenges related to the value of imported components. Power supply is an important component in the ROV, which is used for providing power to the ROV. Meanwhile, in efforts to align with the Indonesian government, it is necessary to consider using domestic components. Seeing the importance of increasing the value of TKDN in the development and procurement of domestic ROVs in order to support the government's efforts in fulfilling the 2015 - 2035 National Industrial Development Master Plan Goals and Targets for Increasing the Use of Domestic Production (P3DN) [4], it is necessary to conduct research and strategies to analyze components for ROV that can be built and produced domestically.

2 Methodology

The methodology in this research refers to the assessment of product local content. The steps in this research can be seen in Fig. 1. First, making the ROV power supply breakdown structure part. Next, a survey was carried out to determine the local component value for each component that makes up the power supply in accordance with the values registered in the data from the Ministry of Industry of the Republic of Indonesia. The data that has been obtained is then processed in accordance with the Minister of Industry Regulation No. 31 of 2022 [5]. The processing results are then added to other overhead costs to obtain the final TKDN value. This final result will be evaluated if the TKDN value is less than satisfactory, then the process returns to calculating the TKDN value.

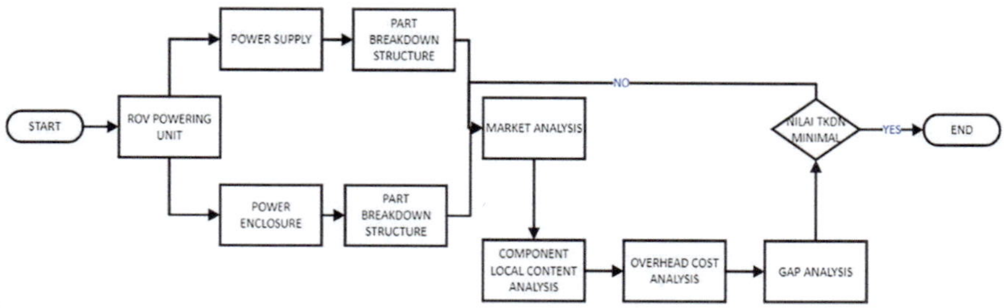

Fig. 1 Flowchart methodology

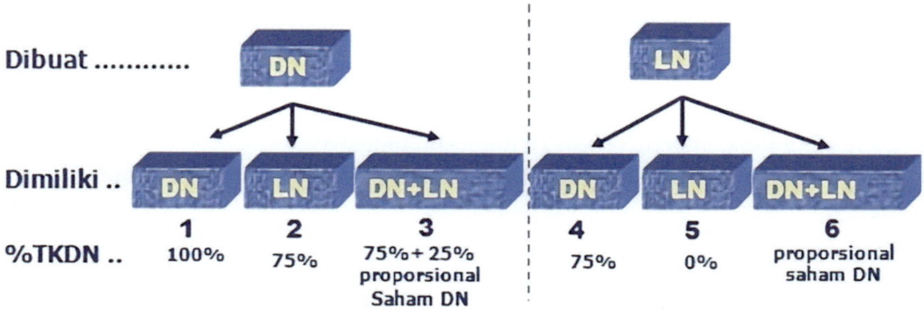

Fig. 2 TKDN according to ministry of industries [4]

2.1 TKDN

TKDN is the amount of the domestic component of goods and/or services expressed in percentage form. In accordance with Presidential Regulation Number 54 of 2010 concerning government procurement of goods or services, the TKDN value plus the required company benefit weight (BMP) is not less than 40%. BMP is a value of appreciation to companies that invest in Indonesia because it is a value of appreciation to companies that invest in Indonesia for empowering micro and small businesses as well as small cooperatives through partnerships; maintaining health, work safety and the environment; empowering the environment (community development); as well as providing after-sales service facilities [6]. Regarding the provisions for determining the TKDN value according to the Ministry of Industry, refer to Fig. 2.

2.2 TKDN Regulation

The procedure for processing TKDN certification in Indonesia has several legal bases, along with a brief explanation as follows:

- Law No. 3 Article 86 of 2014 concerning Industry regarding the obligation to use domestic products in every procurement of goods or services.
- Article 61 Government Regulation 29 of 2018 concerning Industrial Empowerment with the obligation to use domestic products with a TKDN value of at least 25% if there are domestic products with the sum of the TKDN value and BMP Company Benefit Weight of at least 40%.
- Regulation of the President of the Republic of Indonesia Article 66 No. 12 of 2021 concerning government procurement of goods or services related to ministries, institutions and regional apparatus must use PDN, including national design and engineering, where government procurement of goods or services must use domestic products.

In the TKDN certification procedure [7], there are benefits and objectives of TKDN, including:

- Reduce dependence on foreign products.
- Increasing the competitiveness of domestic industry.
- Get a TKDN certificate, which is valid for three years and registered on the P3DN website.
- Synchronize with LKPP E-Catalogue.
- Obtain price preferences in the procurement process.

There are several TKDN provisions by the Ministry of Industry (Ministry of Industry) in (https://www.kemenperin.go.id, 2011) 2011 Ministerial Regulation number 16/M-IND/PER/2/2011 explaining, among others:

- The types of TKDN are goods, services, and combinations.
- Company requirements include investment, location, and production.
- Product requirements include materials (local/imported) and labour (Indonesian/WNA), production process (changes including function, form, and properties), domestic production with higher level provisions.

In the provisions related to the concept of calculating TKDN in several aspects regulated in the Minister of Industry Regulation No. 16/M-IND/PER/2/2011 with the following three components:

- Raw materials: acquisition value/price, value/price of cost components including transportation, handling, etc.
- Labor: salary/salary, allowance costs, facilities, overtime, etc.
- Plant and work equipment: owned (asset depreciation value), rented (value or rental price of other production costs).

In the three components above, two classifications are identified, namely:

- Overseas composition (KLN).
- Domestic composition (KDN).

In the division/breakdown of costs that have been carried out, with the identification and classification above, the cost components (breakdown costs) will be obtained

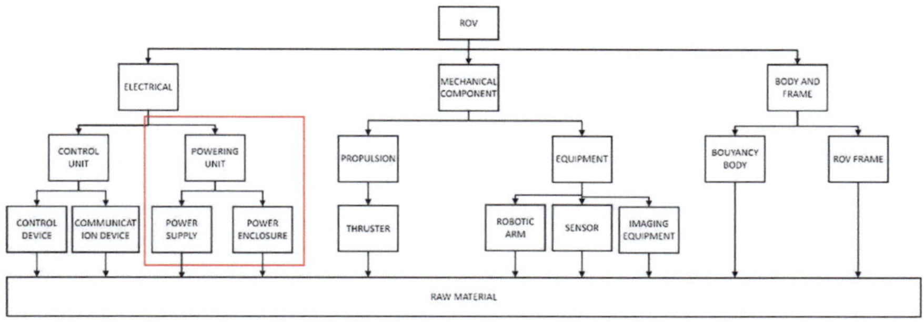

Fig. 3 ROV part breakdown structure

along with their characteristics (variable costs or fixed costs). After identification and classification, the following formula is used to calculate TKDN:

$$\%\text{TKDN} = \frac{\text{Local Content Cost}}{\text{Local Content Cost} + \text{Imported Content Cost}} \times 100\% \tag{1}$$

3 Results and Discussion

3.1 Part Breakdown Structure

Part breakdown structure (PBS) is a hierarchical representation of a product or system that breaks it down into its constituent parts, components, or elements. It is commonly used in project management, engineering, and manufacturing to provide a detailed view of the structure of a product. A PBS is particularly useful for organizing and managing complex projects or products. It can be thought of as a visual tool that helps in understanding the relationship between various parts and sub-parts of a system or product. In this research, the part breakdown structure was carried out twice, first to determine the position of the powering unit in the overall ROV structure as shown in the Fig. 3. Then a further part breakdown structure is made which details the components of the powering unit, which consists of the power supply and power enclosure as well as other constituent components, as can be seen in Fig. 4.

3.2 Market Analysis and TKDN Value Assessment

As a result of the part breakdown structure that has been carried out, the components that make up the powering unit are obtained, which are divided into components:

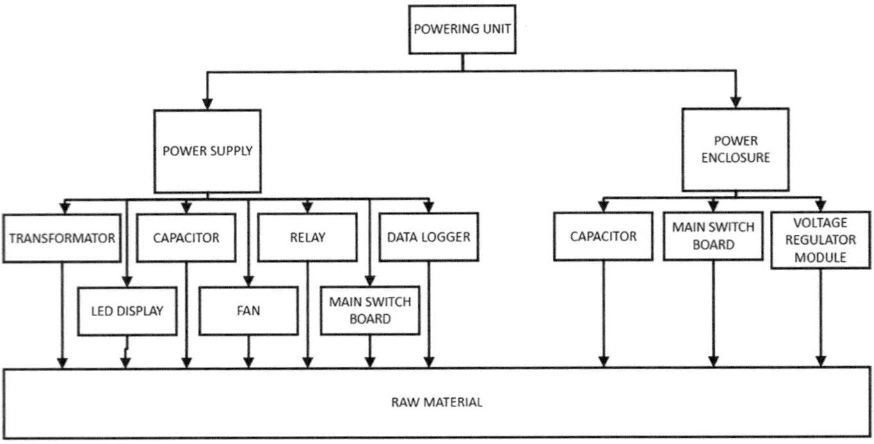

Fig. 4 Powering unit part breakdown structure

- Power supplies
- Power enclosure

Next, an assessment of the TKDN value of each component available on the market is carried out. The TKDN value of each component can be seen in the Ministry of Industry's e-catalogue or via the component's TKDN certificate. Then the price of each component is multiplied by the component's TKDN value so that it is divided into the local component price and the imported component price. The next process is to calculate the overhead costs of manufacturing components, including:

- employee salary costs
- cost of electricity used
- transportation costs
- insurance fee
- testing and trial costs

After calculating the overhead, the TKDN value of the overhead is calculated.

The total TKDN component costs are added together with the overhead TKDN costs to obtain the total TKDN value from manufacturing the powering unit components. The TKDN assessment on the powering unit obtained the following two results:

- The power supply costs a total of Rp. 7,828,550,- with a total TKDN value of 70.0942% as shown in Fig. 5.
- The power enclosure costs a total of IDR 12,522,648 to manufacture, with a total TKDN value of 56,796% as shown in Fig. 6.

Description		ASSESSMENT BY MANUFACTURER			
		Cost per 1 (one) product unit			
		KDN Cost	KLN Cost	Total Cost	TKDN percent
I	Direct Material				
1.	Raw Materials for Direct Materials	1.480.362,65	2.341.187,36	3.821.550,00	18,91
2.	Raw Materials for Other Related Costs	84.500,00	-	84.500,00	1,08
II	Direct Labor				
1.	Direct Labor	2.520.000,00	-	2.520.000,00	32,19
2	Direct Labor for Other Related Costs	-	-	-	-
III	Factory Indirect Costs (*Factory Overhead*)				
1.	Indirect Labor	-	-	-	-
2.	Owned Machines	-	-	-	-
3.	Rental Machines	-	-	-	-
4.	Other Related Indirect Costs	1.402.500,00	-	1.402.500,00	17,92
Production cost		5.487.362,65	2.341.187,36	7.828.550,00	70,09

Fig. 5 Power supply local content value

Description		ASSESSMENT BY MANUFACTURER			
		Cost per 1 (one) product unit			
		KDN Cost	KLN Cost	Total Cost	TKDN percent
I	Direct Material				
1.	Raw Materials for Direct Materials	2.829.863,01	5.410.284,99	8.240.148,00	22,60
2.	Raw Materials for Other Related Costs	360.000,00	-	360.000,00	2,87
II	Direct Labor				
1.	Direct Labor	2.520.000,00	-	2.520.000,00	20,12
2	Direct Labor for Other Related Costs	-	-	-	-
III	Factory Indirect Costs (*Factory Overhead*)				
1.	Indirect Labor	-	-	-	-
2.	Owned Machines	-	-	-	-
3.	Rental Machines	-	-	-	-
4.	Other Related Indirect Costs	1.402.500,00	-	1.402.500,00	11,20
Production cost		7.112.363,01	5.410.284,99	12.522.648,00	56,80

Fig. 6 Power enclosure local content value

4 Conclusion

In summary, this research assesses the TKDN value for the ROV power unit. Based on the results of the TKDN value assessment of the power unit, which consists of the power supply and power enclosure, it was found that the TKDN value was 70.0942% for the power supply and 56.796% for the power enclosure, where this value met the minimum TKDN value qualification of 40%. However, there are still many things that can be done to increase this value. Field findings show that many components, especially complex electrical components, are still imported and should be produced locally.

Acknowledgements This research is used as a graduation requirement and supports the thesis in the master's degree program in Ocean Engineering within the Department of Naval Architecture. It also serves as a collaborative research project with the Department of Naval Architecture at the Institut Teknologi Sepuluh Nopember.

References

1. D. Hartono, W.W. Purwanto, Nurkholis, I.A. Rum, Impact analysis of natural gas policy in Indonesia. Energy Sources, Part B: Econ., Plan. Policy. **12**(8), 699–706 (2017)
2. J. Betancourt, W. Coral, J. Colorado.: An integrated ROV solution for underwater net-cage inspection in fish farms using computer vision. SN Appl. Sci. **2**(12) (2020)
3. M. Fatan, M.R. Daliri, A. Mohammad Shahri, Underwater cable detection in the images using edge classification based on texture information. Meas.: J. Int. Meas. Confed. **91**, 309–317 (2016)
4. kemenperin.go.id, Peningkatan Penggunaan Produksi Dalam Negeri (P3DN). In Kemenperin.go.id (p. 1) (2011). https://www.kemenperin.go.id/artikel/23/Peningkatan-Penggunaan-Produksi-Dalam-Negeri-(P3DN)
5. P3DN, P.D.D.I.K.P, P3DN home. PUSAT dAta dan informasi kementerian perindustrian (2021). http://www.tkdn.kemenperin.go.id/. Accessed 15 Jan 2024
6. PERPRES No 54 tahun 2010, Pub. L. No. 54 tahun 2010, 1 (2010)
7. PT. Surveyor Indonesia, Prosedur Pengurusan Sertifikasi TKDN (n.d.). www.idsurvey.id. Accessed 15 Jan 2024

Agile Manufacturing Practices in the Shipbuilding Industry: Linkage Between Fluctuating Demand, Customer Focus, and Shipyard Performance

Sutrisno Sutrisno, Mokh Suef, and Buana Ma'ruf

Abstract Nowadays, the shipbuilding industry faces a volatile market because global shipyard capacity exceeds demand. The shipyard's management carried out various strategies and approaches to improve operational performance amid fluctuating markets, but many shipyards still have difficulties improving the shipyard performance. Meanwhile, agile manufacturing (AM) is one of the solutions to unpredictable problems, dynamic demand, and products with a high level of customization, which has been widely implemented in the manufacturing industry. Therefore, the present study aims to investigate the linkage between fluctuating demand, customer focus, and shipyard performance from the perspective of AM practices. This study used a quantitative PLS-SEM method and cluster random sampling technique. The research data were collected through a questionnaire survey of 155 respondents from 27 shipyards in Indonesia. The results revealed that fluctuating demand positively influences customer focus, and customer focus positively and significantly impacts shipyard performance in AM practices. This study is one of the few empirical studies on AM practices in the shipbuilding industry. Besides, this study provides insight for shipyard practitioners to consider implementing the AM strategy by enhancing customer focus to increase operational performance amid fluctuating markets.

Keywords Agile manufacturing · Customer focus · Fluctuating markets · Shipbuilding industry · Shipyard performance

---

S. Sutrisno (✉)
Interdisciplinary School of Management and Technology, Institut Teknologi Sepuluh Nopember, Surabaya, Indonesia
e-mail: trisnogrs@gmail.com

M. Suef
Department of Industrial and System Engineering, Institut Teknologi Sepuluh Nopember, Surabaya, Indonesia
e-mail: m_suef@ie.its.ac.id

B. Ma'ruf
The National Research and Innovation Agency, Jakarta, Indonesia
e-mail: buana.maruf@brin.go.id

R. Sjarief Widjaja et al. (eds.), *Marine Technology 4*,
SpringerBriefs in Applied Sciences and Technology,
https://doi.org/10.1007/978-3-032-08840-6_6

1 Introduction

Global business competition is a challenge for industries in various sectors today. Likewise, the shipbuilding industry faces market competition triggered by an imbalance between demand and supply due to the shipyard's capacity exceeding customer demand [1]. Therefore, the shipyard must respond quickly to fluctuating markets [2]. Differentiation in providing excellent service, including faster response and complaint resolution, can increase profitability [3]. Besides, a customer focus strategy has proven effective in improving the company's operational performance [4].

There is much research on the relationship between customer focus and operational performance from the quality management point of view. However, there is less research on this topic in the shipbuilding industry from AM's perspective. AM is the ability to respond quickly and effectively to market changes driven by customer demand to survive and thrive in an unpredictable environment of competition and change [5]. In AM frameworks, customer focus is one of the crucial factors in improving operational and business performance [6–8]. So, this study aims to test the hypothesis that fluctuating demand positively influences customer focus, and customer focus positively impacts shipyard performance in AM practices.

2 Literature Review and Hypotheses Development

2.1 Fluctuating Demand, Customer Focus, and Shipyard Performance

The world shipbuilding market tends to fluctuate, influenced by several important factors, including the price of modern second-hand ships, buyers' financial liquidity, credit availability, and buyer expectations [9]. Based on the facts, the world shipbuilding industry market experienced fluctuations from 2012 to 2022 [10].

Various concepts of customer focus are found in scientific articles [11]. However, in this study, customer focus is seen from the AM perspective, measured by six critical indicators from the references, see Table 1.

Table 1 Customer focus indicators

No	Indicators	References
1	Product quality	[6, 7, 12–15]
2	Customer response	[7, 13–17]
3	Customer satisfaction	[6, 7, 16, 17]
4	Product life cycle	[6, 13, 14, 18]
5	Long-term relationship	[6, 16, 17, 19]
6	Customer-driven innovation	[6, 16, 17]

The operational performance indicators in this study refer to previous research, namely quality, production cost, and building time [20], coupled with sales revenue as one of the critical indicators in the shipyard's key performance indicators [21].

2.2 *Hypotheses Development*

Previous studies confirmed that turbulent environments positively affect the adoption of AM in the manufacturing industry [19]. In addition, the critical variables of AM practices in the shipbuilding industry were identified, and one of the critical variables is customer focus [22]. Thus, the first hypothesis is as follows:

H1: Fluctuating demand (FD) positively influences customer focus (CF)

There is less research on the relationship between customer focus and shipyard performance from the perspective of AM. So, the second hypothesis is as follows:

H2: Customer focus (CF) positively impacts shipyard performance (SP)

Figure 1 shows the conceptual model and hypothesis of the research.

3 Methodology

This study uses the PLS-SEM quantitative method and cluster random sampling technique. The population of this study is the shipyard industry in Indonesia, with a scope of new building business because the construction of new ships is more complex than ship repairs [23]. The minimum number of samples is calculated using G*Power software [24], resulting in 146 samples (6 predictors, 95% power, and 5% margin of error). A questionnaire survey was distributed to respondents online, and 155 valid data were successfully collected from 27 shipyards (see Table 2).

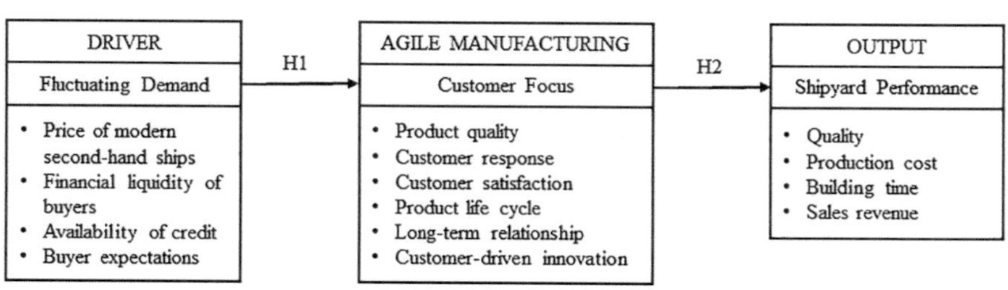

Fig. 1 Conceptual model of research

Table 2 Population and sample

Cluster	Shipyard population		Shipyard scope (new-build)		Sample/data collected			
					Shipyard		Respondent	
	Count	%	Count	%	Count	%	Count	%
Batam	56	24	21	25	7	26	39	25
Outside batam	179	76	62	75	20	74	116	75
Total	235	100	83	100	27	100	155	100

4 Results and Discussion

4.1 Measurement Model Assessment

Table 3 shows the results of the measurement model test using the SmartPLS 4 software. Meanwhile, the validity discriminant is valid (HTMT value is below 0.9) [25].

Table 3 Construct reliability and convergent validity

Constructs and items		Loadings (>0.60) [26]	VIF (<5) [27]	Cronbach's alpha (≥0.70) [28]	Composite reliability (≥0.70) [28]	AVE (≥0.50) [28]
FD	FD1	0.744	1.488	0.750	0.842	0.572
	FD2	0.804	1.588			
	FD3	0.777	1.487			
	FD4	0.696	1.275			
CF	CF1	0.807	2.350	0.905	0.927	0.680
	CF2	0.901	4.003			
	CF3	0.856	2.752			
	CF4	0.726	1.674			
	CF5	0.851	2.991			
	CF6	0.795	2.452			
SP	SP1	0.805	1.463	0.798	0.866	0.619
	SP2	0.774	1.715			
	SP3	0.865	2.370			
	SP4	0.693	1.524			

Table 4 Hypothesis test

Hypothesis	Linkage	Original sample	Standard deviation	T statistics	P-values	F-square
H1	FD- > CF	0.214	0.078	2.752	0.006	0.048
H2	CF- > SP	0.582	0.062	9.454	0.000	0.511

4.2 Structural Model Assessment

The results of structural model testing are shown in Table 4, where the p-values of (FD- > CF) and (CF- > SP) are 0.006 and 0.000, respectively, smaller than the alpha value of 0.050. Therefore, the hypothesis that FD positively influences CF and CF significantly impacts SP is supported. In addition, the path coefficient of (FD- > CF) = 0.214 > 0.15 indicates a moderate relationship. Meanwhile, the path coefficient of (CF- > SP) = 0.582 > 0.35 indicates a significant relationship [28]. The robustness testing of the model was carried out using R-square and f-square. The results show that the R-square of CF and SP is 0.046 and 0.338, respectively. At the same time, the f-square of (FD- > CF) and (CF- > SP) is 0.048 and 0.511, respectively. In addition, the model's predictive power was tested by PLS Predict, where PLS-SEM RMSE and PLS-SEM MAE values are mostly less than LM RSME and LM MAE values, indicating that the model has a medium power prediction [29].

4.3 Discussion

This research results reveal the sequence of the fluctuation demand variables: buyers' financial liquidity, availability of credit, price of modern second-hand ships, and buyer expectations. The validated model of linkage between fluctuating demand, customer focus, and shipyard performance is shown in Fig. 2.

Customer response is the most critical factor that improves operational performance. Increasing customer response speed through comprehensive planning

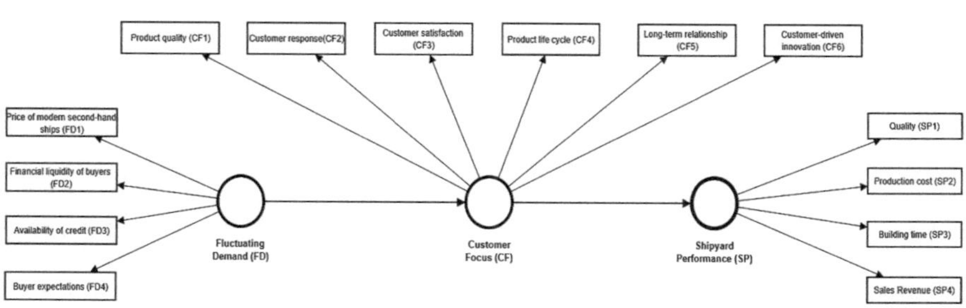

Fig. 2 A validated model of linkage between fluctuating demand, customer focus, and shipyard performance

is needed for competitiveness amid increasing global market competition [30]. Customer satisfaction results from a congruence between perceived product characteristics and customer expectations [11]. Therefore, good products and satisfied service during construction are essential in creating customer loyalty. Meanwhile, long-term customer relationships are essential factors affecting a shipyard's operational performance. From an AM perspective, a partnership with a customer plays a crucial role in responding to market needs quickly [31].

The product quality of the ships determines the ship's service life and operating costs, which affects the business performance of the ship operator company. Customer satisfaction with product quality will encourage repeat orders. Conversely, poor product quality will negatively impact the company's competitiveness [32]. Customer-driven innovation is an effective strategy to realize products that customers crave because customer needs can be directly used as the topic of innovation programs. Notably, technology innovation is critical today and in the future and positively impacts operating performance, cost, flexibility, and delivery [33].

Ship owners or operators require easy operation, maintenance, and long service life. Therefore, the shipyard must be able to make or choose the ship design that suits customer needs and produce a good-quality ship that has a reliable product life cycle. The product life cycle will decrease over time, so the company needs good product flexibility to remain competitive in a competitive environment [7]. Finally, the sequence of shipyard operational performance indicators is building time, quality, operational cost, and sales revenue. This result is relevant to research conducted on 350 manufacturing companies in Puducherry, India [34].

5 Conclusions

The shipyard's operational performance can be improved by strengthening customer-focused development programs. All critical factors of customer focus must be improved simultaneously. Customer satisfaction must be measured periodically and evaluated for improvement to encourage timely delivery, quality improvement, and cost efficiency. A good level of customer satisfaction will encourage the customer to place repeat orders and impact increasing the company revenue.

This research provides empirical evidence for practitioners to consider implementing AM by enhancing customer-focused programs to improve company performance. At the same time, this study contributes to reinforcing the AM framework appropriate for the shipbuilding industry.

6 Limitations and Future Research

This research data was collected only from the shipyards in Indonesia. Therefore, the study results reflect the condition of shipyards in Indonesia that may be different in other countries. In addition, further research on the relationship between shipyard operational performance and the other six dimensions of predictors is still open.

References

1. Y. Wada, K. Hamada, N. Hirata, Shipbuilding capacity optimization using shipbuilding demand forecasting model. J. Mar. Sci. Technol. **27**(1), 522–540 (2022). https://doi.org/10.1007/s00773-021-00852-8
2. M.I. Abubakar, Q. Wang, Key human factors and their effects on human centered assembly performance. Int. J. Ind. Ergon. **69**, 48–57 (2019). https://doi.org/10.1016/j.ergon.2018.09.009
3. P.C. Mandal, Service excellence in marketing for firms: strategies and initiatives. Int. J. Serv. Sci. Manag. Eng. Technol. **13**(1), 1–22 (2022).https://doi.org/10.4018/IJSSMET.290331
4. O. Niyi Anifowose, M. Ghasemi, B.R. Olaleye, Total quality management and small and medium-sized enterprises' (SMEs) performance: Mediating role of innovation speed. Sustain. **14**(14), 8719 (2022). https://doi.org/10.3390/su14148719
5. A. Gunasekaran, Agile manufacturing: a framework for research and development. Int. J. Prod. Econ. **62**(1), 87–105 (1999). https://doi.org/10.1016/S0925-5273(98)00222-9
6. Y.Y. Yusuf, M. Sarhadi, A. Gunasekaran, Agile manufacturing: The drivers, concepts and attributes. Int. J. Prod. Econ. **62**(1), 33–43 (1999). https://doi.org/10.1016/S0925-5273(98)00219-9
7. R. Dubey, A. Gunasekaran, Agile manufacturing: framework and its empirical validation. Int. J. Adv. Manuf. Technol. **76**(9–12), 2147–2157 (2015). https://doi.org/10.1007/s00170-014-6455-6
8. R. Kumar, K. Singh, S.K. Jain, Development of a framework for agile manufacturing. World J. Sci. Technol. Sustain. Dev. **16**(4), 161–169 (2019). https://doi.org/10.1108/WJSTSD-05-2019-0022
9. K. Gourdon, An analysis of market-distorting factors in shipbuilding. OECD Sci. Technol. Ind. Policy Pap. **67** (2019). https://www.oecd-ilibrary.org/science-and-technology/an-analysis-of-market-distorting-factors-in-shipbuilding_b39ade10-en
10. BRS: Shipping and shipbuilding markets: Annual review. **110** (2023). http://www.brs-paris.com/annual/annual-2013/pdf/annual_review_2013-a.pdf%5Cnhttp://www.brs-paris.com/index.php?page=annualreview
11. A. Islamgaleyev, M. Petrova, G. Kurenkeyeva, S. Shalbayeva, A. Kadirbergenova, Increasing customer focus in metal trading. Entrep. Sustain. **8**(1), 604–617 (2020). https://doi.org/10.9770/jesi.2020.8.1(42)
12. Z. Zhang, H. Sharifi, A methodology for achieving agility in manufacturing organisations. Int. J. Oper. Prod. Manag. **20**(4), 496–513 (2000). https://doi.org/10.1108/01443570010314818
13. G. Ramesh, S.R. Devadasan, Literature review on the agile manufacturing criteria. J. Manuf. Technol. Manag. **18**(2), 182–201 (2007). https://doi.org/10.1108/17410380710722890
14. S.A. Raj, S. Vinodh, Forty criteria based agility assessment using scoring approach in an Indian relays manufacturing organization. J. Eng. Des. Technol. **12**(4), 506–517 (2014). https://doi.org/10.1108/JEDT-10-2010-0065
15. M. Goswami, G. Kumar, An investigation of agile manufacturing enablers in Indian automotive SMEs using structural equation model. Meas. Bus. Excell. **22**(3), 276–291 (2018). https://doi.org/10.1108/MBE-10-2017-0068

16. H. Nejatian, M.H. Zarei, M. Nejati, S.M. Zanjirchi, A hybrid approach to achieve organizational agility: An empirical study of a food company. Benchmarking. **25**(1), 201–234 (2018). https://doi.org/10.1108/BIJ-09-2016-0147
17. R. Kumar, K. Singh, S.K. Jain, An evaluation of agile manufacturing initiatives in the Indian manufacturing industry. Int. J. Qual. Reliab. Manag. **37**(1), 156–187 (2020). https://doi.org/10.1108/IJQRM-12-2018-0353
18. R. Sindhwani, V. Malhotra, A framework to enhance agile manufacturing system: A total interpretive structural modelling (TISM) approach. Benchmarking **24**(2), 467–487 (2017). https://doi.org/10.1108/BIJ-09-2015-0092
19. D. Vázquez-Bustelo, L. Avella, E. Fernández, Agility drivers, enablers and outcomes: Empirical test of an integrated agile manufacturing model. Int. J. Oper. Prod. Manag. **27**(12), 1303–1332 (2007). https://doi.org/10.1108/01443570710835633
20. P. Floriano, T. Lamb, C. Souza, Shipbuilding performance benchmarking. Int. J. Bus. Perform. Manag. **11**(3), 216–235 (2009). https://doi.org/10.1504/IJBPM.2009.024372
21. D. Gavalas, T. Syriopoulos, M. Tsatsaronis, Assessing key performance indicators in the shipbuilding industry; an MCDM approach. Marit. Policy Manag. **49**(4), 463–491 (2022). https://doi.org/10.1080/03088839.2021.1876939
22. S. Sutrisno, M. Suef, B. Ma'ruf, Implementation of agile manufacturing in the shipbuilding industry: Challenge and recommendation. in: *Proceedings of the international conference on industrial engineering and operations management*, Michigan, USA: IEOM Society International, 199–208 (2023). https://doi.org/10.46254/AP03.20220033
23. Ö.U. Senturk, The interaction between the ship repair, ship conversion and shipbuilding industries. OECD J. Gen. Pap. **2010**(3), 7–36 (2011). https://doi.org/10.1787/gen_papers-2010-5kg6z7tm3b42
24. F. Faul, E. Erdfelder, A. Buchner, A.G. Lang, Statistical power analyses using G*Power 3.1: Tests for correlation and regression analyses. Behav. Res. Methods. **41**(4), 1149–1160 (2009). https://doi.org/10.3758/BRM.41.4.1149
25. J. Henseler, C.M. Ringle, M. Sarstedt, A new criterion for assessing discriminant validity in variance-based structural equation modeling. J. Acad. Mark. Sci. **43**(1), 115–135 (2015). https://doi.org/10.1007/s11747-014-0403-8
26. W.W. Chin, Issues and opinion on structural equation modeling. Manag. Inf. Syst. Q. (Society Inf. Manag. Manag. Inf. Syst. Res. Center). **22** (1998). https://typeset.io/journals/management-information-systems-quarterly-125psa3b
27. J.M. Becker, C.M. Ringle, M. Sarstedt, F. Völckner, How collinearity affects mixture regression results. Mark. Lett. **26**(4), 643–659 (2015). https://doi.org/10.1007/s11002-014-9299-9
28. J.F. Hair, J.J. Risher, M. Sarstedt, C.M. Ringle, When to use and how to report the results of PLS-SEM. Eur. Bus. Rev. **31**(1), 2–24 (2019). https://doi.org/10.1108/EBR-11-2018-0203
29. G. Shmueli et al., Predictive model assessment in PLS-SEM: guidelines for using PLSpredict. Eur. J. Mark. **53**(11), 2322–2347 (2019). https://doi.org/10.1108/EJM-02-2019-0189
30. T. Stanivuk, M. Šundov, J.Ž Mikuličić, Mišura, A: Logistical activities in the function of development of the shipbuilding industry. Trans. Marit. Sci. **9**(1), 51–62 (2020). https://doi.org/10.7225/toms.v09.n01.004
31. A. Gunasekaran, Agile manufacturing: Enablers and an implementation framework. Int. J. Prod. Res. **36**(5), 1223–1247 (1998). https://doi.org/10.1080/002075498193291
32. S.A.R. Shah, K.R. Jamaludin, H.H.A. Talib, J. Shah, S.M. Yusof, A. Qadeer, Quality management implementation and their impact on operational performance of Pakistan's food processing SMEs. Int. J. Qual. Res. **16**(3), 669–688 (2022). https://doi.org/10.24874/IJQR16.03-01
33. A.B. Abdallah, A.C. Phan, Y. Matsui, Investigating the effects of managerial and technological innovations on operational performance and customer satisfaction of manufacturing companies. Int. J. Bus. Innov. Res. **10**(2–3), 153–183 (2016). https://doi.org/10.1504/IJBIR.2016.074824
34. M. Prabhu, N. Thangasamy, N.N. Abdullah, Analytical review on competitive priorities for operations under manufacturing firms. J. Ind. Eng. Manag. **13**(1), 38–55 (2020). https://doi.org/10.3926/jiem.2876

Development of Port Area Obstacles and Animations in a Virtual Reality-Based Ship Simulator to Enhance User Experience

Trika Pitana, Sutoyo, Muhammad Faqih Asy-Syafi'iy, Nurhadi Siswantoro, Dwi Priyanta, Hari Prastowo, Muhammad Badrus Zaman, and Dian Prasetyawati

Abstract The Indonesian maritime industry has embraced technology from the Industrial Revolution 4.0, including simulators for learning purposes. The Surabaya Shipping Polytechnic, an official school under the Indonesian Ministry of Transportation, has incorporated various simulators, including ship engines and ship bridges, to support cadet education. This research focuses on utilizing virtual reality technology to improve the ship bridge simulator's educational value and reduce manufacturing costs. Virtual reality offers immersive and interactive experiences, making

T. Pitana (✉) · M. F. Asy-Syafi'iy · N. Siswantoro · D. Priyanta · H. Prastowo · M. B. Zaman
Department of Marine Engineering, Institut Teknologi Sepuluh Nopember (ITS), Surabaya, Indonesia
e-mail: trika@its.ac.id

M. F. Asy-Syafi'iy
e-mail: muhammadfaqih46.mf@gmail.com

N. Siswantoro
e-mail: nurhadi@ne.its.ac.id

D. Priyanta
e-mail: priyanta@its.ac.id

H. Prastowo
e-mail: h-prastowo@its.ac.id

M. B. Zaman
e-mail: druz_zaman@ne.its.ac.id

Sutoyo
Politeknik Pelayaran Surabaya, Surabaya, Indonesia
e-mail: Sutoyo@poltekpel-sby.ac.id

M. F. Asy-Syafi'iy
Department of Maritime Studies, Wismar University, Wismar, Germany

D. Prasetyawati
Universitas Muhammadiyah Surabaya, Surabaya, Indonesia
e-mail: dheyan05@yahoo.com

R. Sjarief Widjaja et al. (eds.), *Marine Technology 4*,
SpringerBriefs in Applied Sciences and Technology,
https://doi.org/10.1007/978-3-032-08840-6_7

it a valuable engineering, design, and military training tool. Previous studies have demonstrated that virtual reality can enhance learning effectiveness and skills development by providing realistic simulations and scenarios. This study aims to further enhance the ship bridge simulator by incorporating obstacles and animations, such as 3D ship models placed within the port area environment, ships passing animations, and ship collision animations. Alpha testing or black box testing of the developed application achieved a 100% success rate in functionality testing, while beta testing using the Likert scale method involving 22 cadets yielded excellent ratings in assessing 3D ship models, ships passing animation, and ship collision animations. However, virtual reality may induce certain side effects on health, including oculomotor symptoms, disorientation, and nausea. The study found that a small percentage of cadets experienced little symptoms, such as eye strain, focus difficulty, blurred vision, headache, dizziness, vertigo, stomach awareness, and increased saliva. Understanding and mitigating these side effects are essential to ensure a safe and optimal user experience.

Keywords Animations · 3D ship models · User experience · Virtual reality

1 Introduction

The Indonesian maritime industry has started to make use of technology from the Industrial Revolution 4.0, including the use of drones for inspection, the creation of navigation and communication tools, the vessel traffic service (VTS), the use of augmented reality (AR) and virtual reality (VR) on ships, and the use of simulators to aid in learning.

A simulator is one of many current methods of education or learning that we can use to learn more about something. It is impossible to separate the use of simulators from Surabaya Shipping Polytechnic, one of the official polytechnics run by the Indonesian Ministry of Transportation that has produced numerous trained cadets. At Surabaya Shipping Polytechnic, various simulators have been installed, including simulators for ship engines, ship bridges, and other navigational equipment.

Using simulators as a teaching tool supports cadets' learning, and it will undoubtedly be simpler for cadets to sail in the real world. The development of the potential to use virtual reality technology to apply ship bridge simulators as an educational tool for cadets is what will be used in this case as unique research material. The bridge simulator is intended to reduce the cost of manufacturing the simulator itself and, at this particular time, to aid cadets in learning [1].

Virtual reality is a technology that uses a computer system to simulate and visualize environmental conditions while also involving the user's senses and emphasizing interaction. Through software, virtual reality can create an artificial environment (software). With immersive and interactive experiences in various fields, including science and technology, foreign languages, and social sciences, virtual reality has long held promise as a tool to improve education [2].

Virtual reality technology, including cadets, has been increasingly used in education to support students' learning activities. Virtual reality allows for immersive and interactive experiences, making it a valuable tool for training and learning in engineering, design, and military training. Research studies have shown that VR can enhance learning effectiveness by providing realistic simulations and scenarios that allow students to apply their knowledge and skills in a safe and controlled environment [3]. In military training, VR has been used to simulate combat scenarios, allowing cadets to practice decision-making and tactical skills in a safe and realistic environment [4].

In addition to its use in military training, virtual reality has also been used in engineering and design education to support learning technical skills such as 3D modeling and CAD. Studies have shown that VR can improve students' ability to visualize and manipulate 3D objects and their understanding of spatial relationships and geometric concepts. Virtual reality technology has also been used to support the learning of soft skills such as teamwork, communication, and leadership [5]. Virtual reality simulations can create virtual environments where cadets can practice and develop these skills in a safe and controlled environment.

Tanjung Priok, Tanjung Perak, Tanjung Emas, and Soekarno-Hatta ports are just a few Indonesian ports where 3D development was previously done in the shipping area. The ship bridge simulator components in the Surabaya Shipping Polytechnic are also being developed. Blender software is used to create 3D models, and the Unity 3D software is used to develop packages for virtual reality applications [6]. A Likert scale validation test was run based on the development results to determine how well this application can help cadets gain experience and knowledge following the sailing's original environmental conditions [7][8]. The validation's findings led to recommendations for further development of simulators that can be more useful and approach shipping situations and conditions in line with those in the real world, such as including variables and new features in the ship bridge simulator instrument. The previous generation has developed this ship bridge simulator to be more effective and approach shipping situations and conditions as those in the real world. They have added several new features, such as weather, wave conditions, and shipping channel features.

In addition, several shipboard features must be improved in how well they perform their intended roles to reduce future bugs when using virtual reality. In order to make the shipping conditions in the application as realistic as possible, this study will further develop the ship bridge simulator by adding obstacles and animations, such as variations in the 3D ship models that will be placed around the port area environment. The 3D ship models are also used to develop ship passing animation and ship collision animation, so cadets can have the best immersive experience.

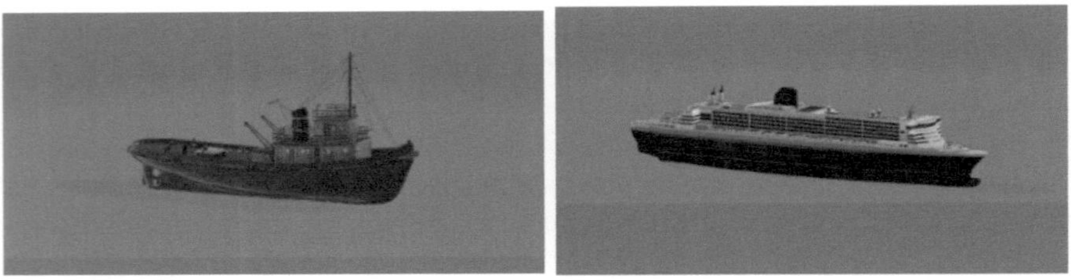

(a) (b)

Fig. 1 3D model of ship **a** tugboat, **b** passenger ship

2 Methodology

The research method is a systematic approach to gathering data that aligns with research goals. It involves various stages, including reviewing existing literature, defining the research problem, collecting and analyzing data, designing and developing virtual reality objects, conducting tests, validating findings, and obtaining results.

2.1 3D Modelling of Ships and Development of Animations

This section explores the process of creating ship models using Blender and developing ship passing and ship collision animations using the Unity software. It then delves into the step-by-step process of creating ship models in Blender, including hull shapes, deck details, and intricate components. The section further explores the integration of these ship models into the Unity software, focusing on developing ship passing and ship collision animations. It discusses techniques such as keyframe animation, particle systems, and physics simulations to bring the ship's movements and collision effects to life. By combining the capabilities of Blender and Unity, this section provides a comprehensive guide for creating realistic ship models and captivating ship animation sequences for immersive simulations, as shown in Fig. 1.

2.2 Testing and Troubleshooting

This section focuses on a virtual reality application's evaluation and quality assessment. The section introduces the black box method, which involves testing the application without knowing its internal workings. It discusses the importance of this method in identifying potential issues, bugs, and inconsistencies within the application's functionality by checking the application using the functional system table.

Additionally, the section explores the Likert scale method, which employs a standardized questionnaire to gather user feedback and assess their satisfaction levels with the virtual reality experience by testing the cadets of Surabaya Shipping Polytechnic using the questionnaire. In order to assess the potential side effects of prolonged virtual reality (VR) usage, a commonly employed method known as the Kennedy simulator sickness questionnaire was utilized. The participants were immersed in a VR environment for an extended period while their experiences and sensations were closely monitored and recorded.

3 Results and Discussion

3.1 3D Model of Ships

This section presents selected ship figures, which are used as a guide in making 3D model designs. The selected ships are made into a 3D design model using the Blender application.

The ship 3D model created using Blender showcases accurate hull contours, intricate details, and realistic texturing, resulting in a visually stunning and highly realistic representation of the ship. Figure 1 shows the result of the 3D model for each type of ship.

3.2 Ships Passing Animation

The development of ships passing animation in the Unity 3D software involves several steps. First, import the 3D ship models into the Unity project, ensuring it has the necessary animations for movement and other actions. Next, set up the scene by placing the ship model in the desired location and adjusting its position, scale, and rotation. Define the path for the ships by creating a series of waypoints or keyframes along the trajectory, forming a realistic and dynamic movement pattern. To animate the ship, create an animation controller or modify an existing one, setting up animation states such as "Idle", "Moving Forward", and "Turning". Utilize animation clips to define the ship's movements, specifying keyframes and transitions for actions like sailing forward, turning, or any other desired motions. Figure 2 shows the result of the ships passing animation.

(a) (b)

Fig. 2 **a b** Ship passing animation results

3.3 Ship Collision Animation

The development of ship collision animation in the Unity 3D software involves several steps. First step, import the ship models into our Unity project and make sure they have Collider components attached. Collider components define the shape and size of the ships for collision detection. Next, set up the scene by placing the ships in their initial positions, considering the trajectory of their collision. Adjust their positions, scales, and rotations as needed. Use Unity's physics engine to simulate realistic collision behavior. Attach rigidbody components to the ships to enable physics interactions. Test the animation to ensure the ships collide convincingly and the effects associated with the collision are visually appealing and immersive. Using Unity's Collider component and physics engine, we can create a captivating animation of ships colliding, bringing realism and impact to the scene, see Fig. 3.

(a) (b)

Fig. 3 **a b** Ship collision animation results

3.4 Functionality Testing Using Black Box Testing Method

After the application development stage is complete, the next step is to conduct the application testing stage. This test aims to evaluate the functionality of the application. Testing is carried out using the black box method, where evaluation is carried out based on the function of each feature in the application, in meeting the criteria and development objectives. At the testing stage, three functionalities were tested at each available port: Tanjung Priok, Tanjung Perak, Tanjung Emas, and Soekarno-Hatta (Makassar) ports.

In each test, several scenarios are implemented to assess the functionality of the features developed in this study. The test results using the black box method show excellent results because every feature tested can function adequately on every existing port.

3.5 Application Assessment Using the Likert Scale Method

After testing the application functionally using the black box method, the application is then tested by user impression on the experience of using a ship bridge simulator. This impression test uses the Likert scale method with the types of statements and the contents of the statements listed in Table 1. This test was carried out on July 3–4, 2023, using a questionnaire containing statements from 22 cadets of Surabaya Shipping Polytechnic. In this test, the assessment is divided into five points, namely: Score 1: Strongly Disagree, Score 2: Disagree, Score 3: Neutral, Score 4: Agree, Score 5: Strongly Agree.

Based on the analysis of the cadet assessment questionnaire consisting of 15 questions and 22 respondents using Likert scale obtained results as follows: (1) the 3D Model of ships assessment field is included in the excellent category because it has a total value of 503, which is in the range of 440–550 and has a percentage value of 91%, (2) the ships passing animation assessment field is included in the excellent category because it has a total value of 496, which is in the range of 440 to 550 and has a percentage value of 90%, (3) the ship collision animation assessment field is included in the excellent category because it has a total value of 489, which is in the range of 440 to 550 and has a percentage value of 89%.

Table 1 Assessment category for each fevelopment fields

Category	Description	Score
Excellent	Quartile III $\leq$ x $\leq$ Maximum Value	440–550
Good	Quartile II (Median) $\leq$ x $\leq$ Quartile III	330–440
Acceptable	Quartile I $\leq$ x $\leq$ Quartile II (Median)	220–330
Poor	Minimum value $\leq$ x $\leq$ Quartile I	110–220

3.6 Testing the Side Effects of Using Virtual Reality

This test was carried out to determine the side effects of using virtual reality technology using the application developed in this study, namely the ship bridge simulator application. The number of testers was 13 cadets of Surabaya Shipping Polytechnic, with a duration of 5–10 min for each person. The categories of this assessment are divided into three, namely oculomotor, disorientation, and nausea. Then it is assessed based on the level of agreement corresponding to the severity of the symptoms felt by the examiner.

Oculomotor

In the classification of oculomotor assessment, there are four symptoms: eye strain, focus difficulty, blurred vision, and headache. Based on oculomotor assessment, in the first symptom (eye strain), 62% of cadets did not feel the effects of eye strain, and 38% felt slight effects. In the second symptom (focus difficulty), 62% of cadets did not feel the effect of focus difficulty, and 38% felt a little effect. In the third symptom (blurred vision), 85% of cadets did not feel the blurred vision effect, and 15% felt a slightly blurred vision effect. In the fourth symptom (headache), 77% of cadets did not feel the effect of a headache, and 23% felt a slight effect. In this Ocumolotor classification assessment, no cadets felt the symptoms at moderate and severe levels of symptoms.

Disorientation

In the classification of disorientation assessment, there are two symptoms: dizziness and vertigo. Based on disorientation assessment, in the fifth symptom (dizzy), 23% felt a slight dizzy effect. In the sixth symptom (Vertigo), 23% felt slight vertigo effects. In this disorientation classification assessment, no cadets felt the symptoms at moderate or severe levels of symptoms.

Nausea

In the classification of nausea assessment, there are three symptoms: stomach awareness, increased salivation, and burping. Based on the nausea assessment, in the seventh symptom (stomach awareness), 8% feel a slight effect of stomach awareness. In the eighth symptom (increased saliva), 15% slightly felt the effect of increased saliva. In the ninth symptom (burp), 100% of cadets do not feel the effect of the burp. In this nausea classification assessment, no cadets felt the symptoms at moderate and severe levels of symptoms.

4 Conclusion

The conclusions that can be obtained from this research are as follows:

(1) Functionality testing (alpha testing) has been carried out on the bridge simulator development application using the black box method. The black box method is divided into three tests, each with its scenario. Black box testing at all ports obtained a success percentage of 100%. All of the tested features can run very well without any trouble.
(2) The second test (beta testing) tested the application on 22 cadets of Surabaya Shipping Polytechnic. After trying the application, the cadets filled out 15 questionnaire questions by dividing them into three assessment fields, namely, 3D models of ships, ships passing animation, and ship collision animation. Furthermore, the data is processed using a Likert scale method. The 3D model of ships assessment field obtains 503 points or 91% (excellent category). Ships passing animation assessment field get a total point of 496 or 90% (excellent category). The ship collision animation assessment field gets points of 489 or 89% (excellent category). All questions get a total point of 1488 or 90% (excellent category).
(3) Using VR can cause side effects on health. Side effects are classified as oculomotor, disorientation, and nausea. Each classification has different symptoms. In the oculomotor classification, 38% (5 students) experienced a little level of eye strain, 38% (5 students) experienced a little level of focus difficulty, 15% (2 students) experienced a little level of blurred vision, and 23% (3 students) experienced a little level of headache. In the disorientation classification, 23% (3 students) experienced little dizziness, and 23% (3 students) experienced little vertigo. In the classification of nausea, 8% (1 student) experienced little stomach awareness, and 15% (2 students) experienced little increased saliva. The rest of the students did not feel any symptoms.

Acknowledgements The authors would like to thank the Department of Marine Engineering, Faculty of Marine Technology, Institut Teknologi Sepuluh Nopember, especially for the Digital Marine Operation and Maintenance Laboratorium, who have provided many supporting facilities and funds to continue this research.

References

1. A. Meilinger, I. Torok, The importance of friction stir welding tool. Prod. Process. Syst. **6**(1), 25–34 (2013)
2. A. Grabowski, Practical skills training in enclosure fires: An experimental study with cadets and firefighters using CAVE and HMD-based virtual training simulators. Fire. Saf. J. **125** (2021). https://doi.org/10.1016/j.firesaf.2021.103440
3. L. Jensen, F. Konradsen, A review of the use of virtual reality head-mounted displays in education and training. Edu. Inf. Tech. **23**(4), 1515–1529 (2018)

4. H. Prastowo, T. Pitana, G. Martha, Components acquaintance of fuel oil system using virtual reality application. I. J. Mar. Eng. Innov. Resch. **5**(4) (2020)
5. Ifanov, P. Jessica, S. Salim, M. Syahputra, P. Suri, A systematic literature review on implementation of virtual reality for learning. Proc. Comp. Sci. 260–265 (2023)
6. S. Chiquet, C. Martarelli, D. Weibel, F. Mast, Learning by teaching in immersive virtual reality–Absorption tendency increases learning outcomes. Learn. Instruc. (2023). https://doi.org/10.1016/j.learninstruc.2022.101716
7. N. Siswantoro, D. Haryanto, M.A. Hikmahwan, T. Pitana, Implementation of metaverse for modeling the ship bridge simulator based on virtual reality as educational purposes, case study: Tanjung Priok Port Area. IOP Conf. Ser.: Earth Environ. Sci. **1166**, 012049 (2023)
8. T. Pitana, M. Pramudhita, N. Siswantoro, J.R. Napitupulu, Application of virtual reality as an alternative training media as complement of ship bridge simulator (case study: Port of tanjung emas semarang). IOP Conf. Ser.: Earth Environ. Sci. **1166**, 012049 (2023)

The Challenges and Opportunities of Blockchain Implementation in the Indonesia Shipping Industry: A Literature Review

Gilang Prakasa, Shantika Dita Putri, and Setyo Nugroho

Abstract The influence of Industry 4.0 is across various sectors, including the shipping industry. The concept of "shipping 4.0" refers to the use of technology to support the development of a brand-new, intricately integrated maritime environment to realize the future transformation of the maritime industry. One critical step is integrating blockchain technology into shipping systems, a prominent aspect of Industry 4.0 applications. In Indonesia, the progression towards shipping 4.0 remains a new phase and is constrained due to resource limitations and recent technological advancements. Applying shipping 4.0 with blockchain technology presents a promising prospect for supporting Indonesia's shipping industry. Through previous literature and research insights, this research aims to shed light on the prospects and challenges of implementing blockchain within the maritime sector. This research seeks to advance the knowledge concerning blockchain development within Indonesia's shipping sector and enhance the nation's journey toward shipping 4.0.

Keywords Industry 4.0 · Shipping 4.0 · Blockchain · Maritime industry

G. Prakasa (✉) · S. D. Putri · S. Nugroho
Department of Marine Transportation Engineering, Sepuluh Nopember Institute of Technology, Kampus ITS Keputih Sukolilo, Surabaya, Indonesia
e-mail: prakasagilang@gmail.com

S. D. Putri
e-mail: shantikadp@gmail.com

S. Nugroho
e-mail: snugroho@seatrans.its.ac.id

R. Sjarief Widjaja et al. (eds.), *Marine Technology 4*,
SpringerBriefs in Applied Sciences and Technology,
https://doi.org/10.1007/978-3-032-08840-6_8

1 Introduction

Industry 4.0 is the fourth industrial revolution that aims to create cyber-physical production systems. It enables real-time operation and decision-making by achieving seamless interoperability between systems, people and the environment [1]. Digitalization and Industry 4.0 are closely related concepts transforming the industrial and economic spheres of life, and digital transformation is regarded as the most significant impact of Industry 4.0 [2]. Industry 4.0 involves the digital transformation of industrial and consumer markets, from the advent of smart technologies to the automation of business activities [3].

The COVID-19 pandemic has triggered "shipping 4.0", the future of maritime digitalization, which is changing all industrial sectors, including the shipping sector. The shipping industry is embracing digitalization to increase cost-efficiency, raise competitiveness, and meet the needs of consumers [4], and shipping 4.0 can improve the ability of shipping-related enterprises to respond to market trends and reduce the operating costs of all business processes [5].

Furthermore, shipping 4.0 is a revolutionary set of interdisciplinary technologies, including microelectronics, sensor technology, data engineering, and security [6]. Blockchain is one of the technologies that has the potential to do it by improving supply chain transparency, traceability, and efficiency.

From many perspectives, this study aims to discover the potential for blockchain implementation in maritime transport and how Indonesia's shipping industry could use it.

2 Methodology

This study begins by looking for sources about "Industry 4.0", "shipping 4.0", "digitalization", and "blockchain technology" and their relevance in academic social networking sites for researchers. The journal resources from those sites are then analyzed to determine the topics' significance, analyze the discussion, and draw conclusions.

Five parts of the study structure are set out below. The introduction and methodology explain "why" and "how" this study is doing. The findings will discuss the concepts of this study, and the answers to the objectives of this study will be addressed. The results and future work will be summarized and suggested to continue the study.

3 Findings

3.1 Shipping 4.0

Shipping 4.0 is the maritime industry's future, integrating digital tools, workflows, and approaches into every aspect of maritime business, driven by the fourth industrial revolution [1]. Reza Emad and Shahbakhsh [7] state that shipping has evolved gradually from traditional to modern, technology-driven practices. Shipping 3.0 brought automation and computerized systems onboard, leading to shipping 4.0's goal of achieving full autonomy through advanced technologies. Shipping 4.0 has radically transformed human–machine interactions and shipping operations.

Some of the technologies that become a core of the development of shipping 4.0 include the internet of things (IoT), blockchain, artificial intelligence (AI), and automation [7]:

- Internet of Things

IoT has a role in enabling the digital transformation of the shipping industry by integrating advanced technologies into ship operation and maintenance [8].

- Blockchain

Blockchain is transforming the maritime industry by enabling real-time operations, reducing costs, and improving safety. But it requires changes in operating models and cybersecurity measures [9].

- Artificial intelligence

AI empowers digital transformation in shipping via advanced tech in ship operation and maintenance. It automates tasks, drives digital transformation, and enhances safety and efficiency through real-time operations [10].

- Automation

Autonomous ships can improve the shipping industry's sustainability, resilience, and safety while reducing lifetime costs [11].

With all of the opportunities in technology development, implementing shipping 4.0 initiatives can also present several challenges and barriers for shipping companies, such as [12]:

- The maritime industry's logistics systems and processes can be challenging due to their complexity.
- Shipping companies may lack the necessary resources, including skilled personnel, to implement and manage shipping 4.0 technologies effectively.
- Adopting new technologies can be a significant obstacle for companies relying on legacy systems.
- The shipping industry may resist change, especially when new technologies disrupt established practices.

- With the increased use of digital technologies, ensuring data security is a critical challenge. Robust data protection measures are needed to implement shipping 4.0 initiatives successfully.

3.2 Blockchain Concept

Blockchain technology is a concept that refers to a distributed and decentralized system of storing and verifying data transactions securely and transparently. According to Wagner and Wisnicki [13], blockchain technology provides a new way of exchanging value and information online between individuals and companies. Blockchain is a revolutionary technology that serves as a distributed ledger, providing a secure and transparent way to record transactions. It is a decentralized system storing data across multiple computers and nodes rather than in a single location. This distributed nature of blockchain ensures that no single entity controls the entire network, enhancing security and reducing the risk of fraud [14].

Other perspectives from Yli-Huumo et al. [14] state that the term "blockchain" comes from how data is stored. Information is grouped into blocks, each containing a set of transactions. Once a block is filled with data, it is added to the chain of previous blocks, forming a "blockchain". Each block contains a unique hash code created using cryptographic algorithms. This hash is like a digital fingerprint representing the block's data. Blocks are linked together in the blockchain through these hashes. Each block contains the previous block's hash, creating a blockchain. If data in a block is altered, it changes the hash of that block and breaks the chain. Each block in the chain contains the previous block's hash, so changing data in one block would require changing data in all subsequent blocks, which is impractical [14].

Blockchain technology has several potential applications in the shipping industry, including [15]:

- Blockchain automates container management documentation, including customs declarations and bills of lading.
- Blockchain technology is used in the shipping industry to track and authenticate payment transactions.
- Blockchain can help supply chain management by recording the movement of goods and funds, like a traditional ledger. This can reduce shipping delays caused by slow manual processes.
- Blockchain allows authorized users to track orders, accounts, production schedules, and real-time shipping data.
- Blockchain technology can replace intermediaries to establish trust and confirm authenticity in supply chain collaborations.

Implementing blockchain technology in the shipping industry can present several challenges [16, 17]:

- The cost of developing and maintaining blockchain technology can be quite high.

- Transitioning from older to new blockchain-based systems can be expensive and time-consuming.
- This technology consumes a lot of energy, requiring live power to operate in real-time.
- Technology in shipping can reduce paperwork and costs, but may introduce complexities in system integration and data management.

4 Discussion and Implications

4.1 How is Blockchain Implemented in the Indonesian Shipping Industry?

Blockchain technology has been implemented in Indonesia's shipping industry to enhance supply chain transparency, reduce administrative costs, and combat issues like counterfeit goods, compliance violations, delays, and waste [18]. The adoption of blockchain in Indonesia's ports and shipping industry can support digital transformation and improve supply chain management [19].

The Indonesian fishing industry has integrated blockchain technology into its supply chain models. The World Wildlife Fund (WWF) has collaborated with Sea Quest Fiji Ltd to successfully implement blockchain technology, which could potentially facilitate blockchain implementation in the Indonesian fishing industry [18].

Research suggests that Indonesia has the potential to implement blockchain technology across all industries. Larissa and Parung [18] successfully designed a new supply chain model in the fishing industry using blockchain technology. Alamsyah et al. [20] researched the use of blockchain technology in the halal food industry, a topic of great interest in Indonesia due to the population's concern for halal food. Tedjakusuma and Yahya [21] researched this technology's potential in Indonesia's growth in the non-financial sector of micro, small, and medium enterprises (MSMEs).

Research suggests that implementing blockchain technology can help develop shipping industries and businesses in Indonesia [18, 20, 21]:

- The Indonesian government can collaborate with blockchain companies to develop and implement solutions for the shipping industry.
- The government can provide stakeholders in the shipping industry with training and education on how to effectively use blockchain technology.
- The government can incentivize companies to adopt blockchain technology in the shipping industry.
- The government can establish regulations for utilizing blockchain technology in the shipping industry.

All stakeholders, including the government, shipping companies, authorities, and customers, must fully support the strategies for smooth implementation.

4.2 Challenges and Opportunities

With all of the potential, there are obstacles, challenges, and opportunities for implementing blockchain technology in the Indonesian shipping industry [22, 23]:

Challenges:

- Managing complex systems requires careful planning and coordination of stakeholders, considering factors such as technology, organizational structure, and the external environment.
- The Indonesian shipping industry faces challenges in implementing blockchain technology due to the lack of standardization and a common approach to skilling people. Skilled professionals are required to design, develop, and maintain blockchain networks, which may be scarce and expensive.
- There is currently no regulatory framework for implementing blockchain technology in the shipping industry in Indonesia.
- As mentioned in the previous chapter, the development and maintenance costs have become a challenge for stakeholders in the Indonesian shipping industry.
- Scalability issues hinder blockchain adoption due to differing user interests.

Opportunities:

- Implementing this strategy can improve supply chain transparency and traceability and reduce costs.
- The application can promote decentralized decision-making, resulting in a more democratic process, and can help liberate decision-making.
- Technology can ensure supply chain data security through cryptographic algorithms and distributed ledgers, which prevent data tampering and fraud.
- Shipping companies can gain a competitive advantage by implementing this solution, which can help strengthen their position in the market.
- Sustainability measures can boost industry efficiency while reducing carbon emissions, improving food safety, and promoting social inclusion.

4.3 Elaboration on the Ideas for the Implementation of Blockchain in Indonesia's Shipping Industry

Blockchain can benefit Indonesia's shipping industry, given its archipelagic geography. Its implementation can enhance efficiency, transparency, and security, and help transport goods across the vast network of islands.

Blockchain simplifies logistics and reduces transit times and costs. It provides a transparent platform for coordination and real-time tracking of shipments. This is crucial in Indonesia due to its vast archipelago that presents logistical challenges leading to delays, inefficiencies, and increased costs.

Moreover, blockchain can improve SMEs in Indonesia's economy by reducing paperwork, enhancing trust, and facilitating market access. Blockchain-based finance solutions and smart contracts can create climate-resilient supply chains for the maritime industry.

In the end, collaboration between the private sector and the government is essential to implement blockchain in Indonesia's shipping industry successfully. Creating a clear regulatory framework and standards for blockchain adoption can guide industry players, creating a favourable environment for innovation and technology integration.

5 Conclusion and Future Works

This study reviews shipping 4.0 and its development with blockchain in Indonesia. Challenges include high costs and a lack of standardization, regulations, and technology improvements. However, the system offers potential benefits to stakeholders for competitiveness and sustainability worldwide.

Two potential topics for future discussion are implementing the shipping 4.0 framework in Indonesia and using blockchain technology in the Indonesian shipping industry to discuss current trends such as e-commerce growth, the sea toll program, and increased vessel demand.

Acknowledgements The authors thank the Ministry of Transportation Republic of Indonesia for collaborating with Sepuluh Nopember Institute of Technology (ITS) and providing the Magister Double Degree Marine Transportation Engineering ITS—Rotterdam University of Applied Sciences Programme scholarship.

References

1. G, Aiello, A. Giallanza, G. Mascarella, Towards shipping 4.0. A preliminary gap analysis, Procedia Manuf. **42**, 24–29 (2020). https://doi.org/10.1016/j.promfg.2020.02.019
2. M.Z. Yaqub, A. Alsabban, Industry-4.0-enabled digital transformation: prospects, instruments, challenges, and implications for business strategies. Sustainability. **15**(11), 8553 (2023). https://doi.org/10.3390/su15118553
3. M. Ghobakhloo, Industry 4.0, digitization, and opportunities for sustainability. J. Clean. Prod. **252**, 119869 (2020). https://doi.org/10.1016/j.jclepro.2019.119869
4. Y. Ichimura, D. Dalaklis, M. Kitada, A. Christodoulou, Shipping in the era of digitalization: Mapping the future strategic plans of major maritime commercial actors. Digit. Bus. **2**(1), 100022 (2022). https://doi.org/10.1016/j.digbus.2022.100022
5. Y. Yang, T. Gai, M. Cao, Z. Zhang, H. Zhang, J. Wu, Application of group decision making in shipping industry 4.0: Bibliometric analysis, trends, and future directions. Systems. **11**(2), 69 (2023). https://doi.org/10.3390/systems11020069
6. M.A. Lambrou, M. Ota, Shipping 4.0: Technology stack and digital innovation challenges. Paper presented at IAME 2017 Conference, Kyoto, Japan, 27–30 Aug 2017 (2017)
7. R.G. Emad, M. Shahbakhsh, Digitalization transformation and its challenges in shipping operation: The case of seafarer's cognitive human factor. Adv. Transp. **60**, 684–690 (2022). https://doi.org/10.54941/ahfe1002505

8. K. Hiekata, S. Wanaka, T. Mitsuyuki, R. Ueno, R. Wada, B. Moser, Systems analysis for deployment of internet of things (IoT) in the maritime industry. J. Mar. Sci. Technol. **26**(2), 459–469 (2021). https://doi.org/10.1007/s00773-020-00750-5
9. M. Jović, E. Tijan, D. Žgaljić, S. Aksentijević, Improving maritime transport sustainability using blockchain-based information exchange. Sustainability. **12**(21), 8866 (2020). https://doi.org/10.3390/su12218866
10. D. Gavalas, T. Syriopoulos, E. Roumpis, Digital adoption and efficiency in the maritime industry. J. Shipp. Trade. **7**(1), 11 (2022). https://doi.org/10.1186/s41072-022-00111-y
11. G. Theotokatos, J.L.D. Dantas, G. Polychronidi, G. Rentifi, M.M. Colella, Autonomous shipping—An analysis of the maritime stakeholder perspectives. WMU J. Marit. Aff. **22**(1), 5–35 (2023). https://doi.org/10.1007/s13437-022-00290-2
12. M. Cichosz, C.M. Wallenburg, A.M. Knemeyer, Digital transformation at logistics service providers: barriers, success factors and leading practices. Int. J. Logist. Manag. **31**(2), 209–238 (2020). https://doi.org/10.1108/IJLM-08-2019-0229
13. N. Wagner, B. Wisnicki, Application of Blockchain Technology in Maritime Logistics. DIEM: Dubrov. Int. Econ. Meet. **4**(1), 155–164 (2019)
14. J. Yli-Huumo, D. Ko, S. Choi, S. Park, K. Smolander, Where is current research on blockchain technology? A systematic review. PLoS ONE **11**(10), e0163477 (2016). https://doi.org/10.1371/journal.pone.0163477
15. L. Li, H. Zhou, A survey of blockchain with applications in maritime and shipping industry. Inf. Syst. E-Bus. Manag. **19**(3), 789–807 (2021). https://doi.org/10.1007/s10257-020-00480-6
16. S. Tsiulin, K.H. Reinau, O.-P. Hilmola, N. Goryaev, A. Karam, Blockchain-based applications in shipping and port management: A literature review towards defining key conceptual frameworks. Rev. Int. Bus. Strategy. **30**(2), 201–224 (2020). https://doi.org/10.1108/RIBS-04-2019-0051
17. K. Gerakoudi-Ventouri, Review of studies of blockchain technology effects on the shipping industry. J. Shipp. Trade. **7**(1), 2 (2022). https://doi.org/10.1186/s41072-021-00105-2
18. S. Larissa, J. Parung, Designing supply chain models with blockchain technology in the fishing industry in Indonesia. IOP Conf. Ser. Mater. Sci. Eng. **1072**(1), 012020 (2021). https://doi.org/10.1088/1757-899X/1072/1/012020
19. D.H. Alahmadi, F.A. Baothman, M.M. Alrajhi, F.S. Alshahrani, H.Z. Albalawi, Comparative analysis of blockchain technology to support digital transformation in ports and shipping. J. Intell. Syst. **31**(1), 55–69 (2021). https://doi.org/10.1515/jisys-2021-0131
20. A. Alamsyah, N. Hakim, R. Hendayani, Blockchain-based traceability system to support the indonesian halal supply chain ecosystem. Economies. **10**(6), 134 (2022). https://doi.org/10.3390/economies10060134
21. A.P. Tedjakusuma, B.N. Yahya, A study into opportunities and challenges of blockchain adoption for sustainable nonfinancial sector micro, small, medium enterprises–case studies in Indonesia. in *Proceedings of the 17th International Symposium on Management (INSYMA 2020)*, Vung Tau, February 2020. Adv. Eco. Bus. Manag. Res. **115**, 30–35 (2020). https://doi.org/10.2991/aebmr.k.200127.007
22. A. Yadlapalli, S. Rahman, P. Gopal, Blockchain technology implementation challenges in supply chains—evidence from the case studies of multi-stakeholders. Int. J. Logist. Manag. **33**(5), 278–305 (2022). https://doi.org/10.1108/IJLM-02-2021-0086
23. R.M. Díaz, L. Valdés Figueroa, G. Pérez, Blockchain implementation opportunities and challenges in the Latin American and Caribbean logistics sector. FAL Bulletin. **387**(3) (2021)

Application of Multi-attribute Decision-Making to Determine the Preferred Strategy in Reducing the Casualty Risk Involving Indonesia's Domestic Ferries and RoRo Passenger Operation

Aleik Nurwahyudy, Trika Pitana, and Setyo Nugroho

Abstract Domestic RoRo passenger ferry transport can be considered a sensitive transportation mode. The sensitivity of operation is due to the significant role of service as the primary logistic distribution path connecting the islands within the Indonesian region. Based on existing accident data, it is unfortunate that domestic ferry transportation consistently faces accidents of a severe nature. From 2014 until 2022, nearly every year has witnessed accidents resulting in the loss of one or more RoRo and ferry vessels or rendering them inoperable. This study employs ELECTRE, a MADM method, to determine the most preferable strategies for preventing casualties. The criteria and strategy scores are weighed in these methods by collecting survey-based opinions from the relevant respondents. From the analysis conducted through the method, the strategy involving a change in field operation patterns and enhanced supervision in operational field patterns emerges as the primary strategy requiring significant attention for implementation. This mitigation effort concerns reducing the casualties involving domestic RoRo passengers in Indonesia.

Keywords MADM · Domestic RoRo passenger · ELECTRE

A. Nurwahyudy (✉) · T. Pitana
Department of Marine Engineering, Faculty of Marine Technology, Sepuluh Nopember Institute of Technology , Surabaya, Indonesia
e-mail: aleiknurwahyudy@gmail.com

T. Pitana
e-mail: trika@its.ac.id

S. Nugroho
Department of Sea Transportation, Faculty of Marine Technology, Sepuluh Nopember Institute of Technology , Surabaya, Indonesia
e-mail: snugroho@its.ac.id

R. Sjarief Widjaja et al. (eds.), *Marine Technology 4*,
SpringerBriefs in Applied Sciences and Technology,
https://doi.org/10.1007/978-3-032-08840-6_9

1 Introduction

Public transportation services are a complex system that provides transportation modes to the public, with key indicators including safety, security, comfort, and service quality. Therefore, standards for various aspects of public transportation services are regulated in such a way as to ensure a higher level of supervision and operations [1–3]. For archipelagic nations like Indonesia, maritime transportation is crucial in all aspects of national development. It provides connections and open access to all parts of the country. Indonesia's maritime policies and concepts consider the sea not to divide the nation but to connect all islands. From this perspective, maritime transportation is regarded as safeguarding national unity and an essential support for the economic development equalization program [4].

RoRo-type vessels are considered the most suitable mode of transportation for Indonesia due to their flexibility, low tariffs, and affordable technology [1, 5]. The vessel also can access inland waters that require vessels with relatively shallow drafts. On the other hand, coastal ferry crossings in certain regions of Indonesia also significantly save time and enhance interconnectivity between different areas.

Despite robust arrangements on every aspect of ship operation, casualties at any level continue (Fig. 1). The cause of ship total loss varies depending on the nature of the accident, but is mainly related to the ineffectiveness of damaged control [7, 8, 10]. Relevant safety authorities have issued recommendations to prevent reoccurrences [9]. The resolution focus should address the root causes behind the casualty (indirect causes) [2]. By doing so, the solutions offered are expected to be effective by directly addressing the initial triggers of the accident [7, 10]. However, resolution efforts also need to target the stages directly related so that mitigating the direct causes also contributes to preventing accidents. Determining which actions precede others requires additional methodologies, considering other influential factors.

This paper proposes mitigation strategies to reduce casualties, informed by safety issues identified in prior studies and investigations, incorporating insights from entities like IMO and non-government organizations. The strategies are presented in a list, allowing stakeholders to prioritize the most appropriate and timely measures for reducing the likelihood of future casualties.

2 Methodology

2.1 Study Flow and Selection of Analysis Model

The flow of the study covers a range of steps, including the development of focus study, selection of analytical models, identification of safety issues, development of the strategy to prevent re-occurrence, development of criteria to determine the strategy, determining the weight of criteria, determine the value for each strategy

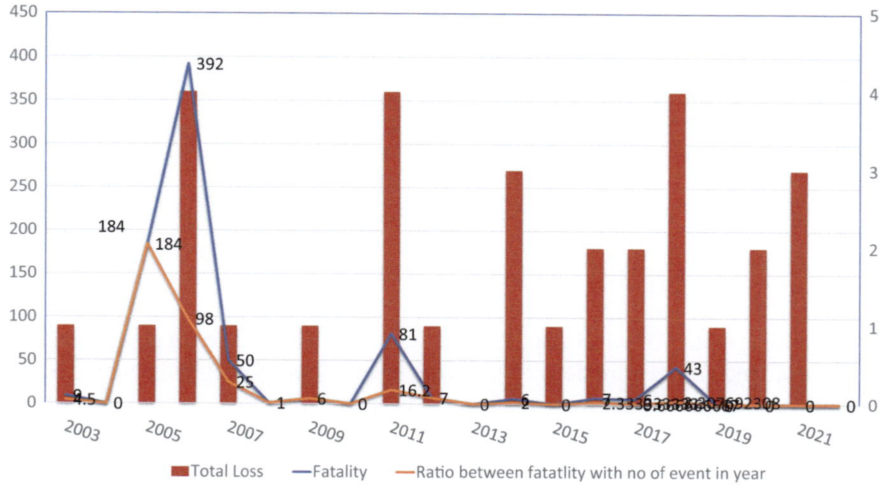

Fig. 1 Data on total loss of Indonesian domestic RoRo pax [14]

based on the criteria defined in previous step, rank and select the strategy based on the highest value as formulated in the model.

This study employs the multi-attribute decision-making (MADM) method to determine the best alternative strategies for preventing accidents involving domestic ferries and RoRo passenger ships. The main objective is to present qualitative and quantitative considerations of the formulated and proposed strategies from different perspectives.

This study utilizes ELECTRE (Elimination et Choix Traduisant la Realité) methods to assess safety improvement strategies in domestic ferry and RoRo transportation. ELECTRE is a technique that combines qualitative and quantitative features to rank and identify the best alternatives, enhancing decision-making abilities [11]. The ELECTRE comprises seven main steps [12].

1. Normalize the decision matrix using the following equation.

$$r_{\text{ij}} = \frac{x_{\text{ij}}}{\sqrt{\sum_{i=1}^{m} x^{2\text{ij}}}} \text{for } I = 1, 2, 3, \ldots, m \text{ and } j = 1, 2, 3, \ldots \tag{1}$$

   A normalized matrix R is developed from the calculation of the formula above.

$$R_{\text{mn}} = \begin{bmatrix} r_{11} & r_{1n} \\ r_{m1} & r_{mn} \end{bmatrix} \tag{2}$$

2. Weigh the normalized decision matrix using the formula:

$$V = R.W \tag{3}$$

3. Develop concordance set and discordance set: For every pair, alternative k and l (where $k.l = 1,2,3,..,m$) are divided into concordance and discordance. Criteria are considered as concordance set when:

$$C_{\mathrm{kl}} = \{j | y_{\mathrm{kj}} \geq y_{\mathrm{lj}}\}; \text{ for } j = 1, 2, 3, \ldots N \tag{4}$$

On the other hand, a complementary called discordance is present when:

$$D_{\mathrm{kl}} = \{j | y_{\mathrm{kj}} < y_{\mathrm{lj}}\}; \text{ for } j = 1, 2, 3, \ldots N \tag{5}$$

4. Develop a matrix of concordance and discordance: The values of elements in the concordance matrix are derived from the summation of weights included in the subset of concordance, with the formula is as follows:

$$C_{\mathrm{kl}} = \sum j \in C_{\mathrm{kl}} W_j; \text{ for } j = 1, 2, 3, \ldots, N \tag{6}$$

The discordance element matrix is derived by dividing the maximum gap value using the following formula:

$$d_{\mathrm{kl}} = \frac{\max\{|y_{\mathrm{kj}} - y_{\mathrm{ij}}|\} j \in D_{\mathrm{kl}}}{\max\{|y_{\mathrm{kj}} - y_{\mathrm{lj}}|\}_{\forall j}} \tag{7}$$

5. Develop a dominance matrix of concordance and discordance: Threshold value c from Eq. (8) is used to develop a concordance matrix and incorporated in the following formula:

$C_{\mathrm{kl}} \geq \underline{c}$ where $\underline{c}$ derived from

$$\underline{c} = \frac{\sum_{k=1}^{m} \sum_{l}^{m} c_{\mathrm{kl}}}{m(m-1)} \tag{8}$$

Matrix element F is determined from the following equation.

$$f_{\mathrm{kl}} = \begin{cases} 1, \text{if} c_{\mathrm{kl}} \geq \underline{c} \\ 0, \text{if} c_{\mathrm{kl}} < \underline{c} \end{cases} \tag{9}$$

The discordance is derived from a matrix G developed using threshold value $\underline{d}$. The $\underline{d}$ value is acquired from the following equation:

$$\underline{d} = \frac{\sum_{k=1}^{m} \sum_{l}^{m} d_{kl}}{m(m-1)} \tag{10}$$

Hence, the matrix G was developed using the equation:

$$g_{kl} = \begin{cases} 1, \text{if} d_{kl} \geq \underline{d} \\ 0, \text{if} d_{kl} < \underline{d} \end{cases} \quad (11)$$

6. Determine the aggregate of the dominance matrix. The formula below is used to develop the matrix.

$$e_{kl} = f_{kl} x g_{kl} \quad (12)$$

7. Eliminate less favourable strategies. Matrix *e* provides the ranking of choices for each alternative, where if $e_{kl} = 1$, then alternative A_k is considered a better choice than A_l.

2.2 Development of Study Questionnaires

Weighting value was performed based on predefined indicators. Similarly, each proposed alternative strategy must be assigned values based on the existing indicators. A survey questionnaire model was developed and distributed to determine the weights for each criterion and the importance of the strategy. The questionnaire, targeting regulatory authorities, ferry operators, academics, and field professionals, gathers perceptions on safety in ferry and domestic RoRo ships. Administered online via Google Forms, responses are visually presented in graphs and spreadsheets. The survey addresses two main aspects: assessing priority levels of safety indicators and gauging opinions on the relationship between proposed strategies and safety indicators. The study received 40 responses, depicted proportionally in Fig. 1.

3 Identification of Safety Issues and Development of Strategy Criterion

3.1 Development of Indicators and Strategies as Well as Their Weight and Value

Indicators used to determine strategy values are based on safety performance criteria from OSHA (Occupational Safety and Health Administration), focusing on improvement efforts viewed through the factors: prevention, reduce consequences, improvement, safety performance optimization, and increased stakeholder participation [13]. From the survey, the weight assigned to each indicator is based on respondents' perceptions of which indicators are more significant compared to others (Fig. 2). In addition, the conception above is translated into the criteria as stated in Table 1.

Table 1 Criteria and their weight for evaluating each proposed strategy

Indicator	Criteria	Weight (%)
C1	Prevention of re-occurrence	36.5
C2	Reduction of cost/ consequences	15.3
C3	Improving productivity rate and organizational performance as a whole	16.5
C4	Optimizing the performance of safety	15.3
C5	Increase stakeholder involvement	16.5

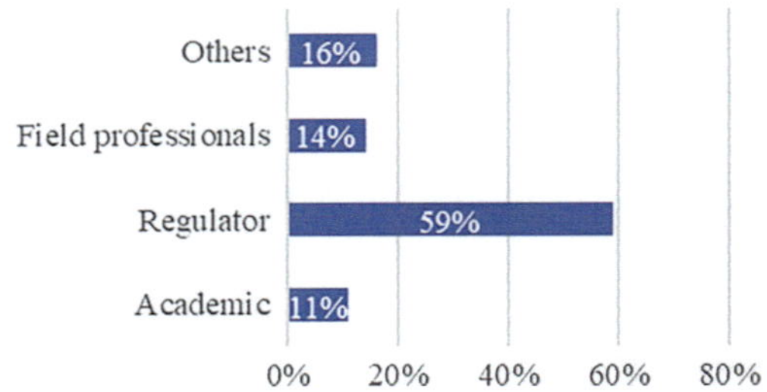

Fig. 2 Proportions of respondents based on their working background

3.2 *Development of Strategies*

The concept of strategy development aims to identify the best options based on the most prominent aspects of the issues, particularly concerning casualty in domestic ferry and RoRo passenger service. Subsequently, these identified issues are compared with KNKT (National Transportation Safety Committee) recommendation data to justify the proposed strategies that offer prospects for addressing safety problems primarily related to high-profile cases [14]. The strategies are evaluated as positive steps in addressing issues related to casualty involving domestic ferries and RoRo passengers. Figure 3 presents a correlation between the findings from different studies and the proposed strategies.

All strategies are initially assumed to have equal urgency for implementation. Each strategy is stated quantitatively to present the trend and be used in the following analysis step. The number in Table 2 is based on the highest percentage of respondents' preference for selecting the strategy under each criterion.

4 Results and Discussion

The ELECTRE model's final results, presented in Table 3, favour strategy 1 (change in operational level work patterns) as the preferred choice, followed by strategy 5. Strategy strengths are indicated by the total tally points, with strategy 1 accumulating 5 points and strategy 2 with 4 points.

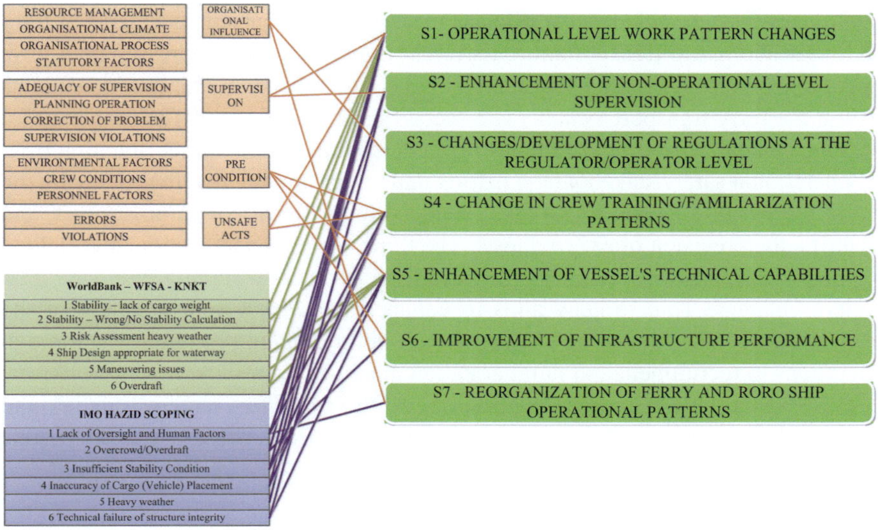

Fig. 3 Correlation of identified safety issues in the domestic RoRo passenger ferries with proposed strategies [8, 9, 14]

Table 2 Strategy value from the questionnaire

Strategy	Indicator (qualitative)				
	C1	C2	C3	C4	C5
	36%	15%	16%	15%	16%
S1	0.67	0.57	0.62	0.80	0.57
S2	0.50	0.50	0.47	0.55	0.40
S3	0.45	0.45	0.50	0.57	0.47
S4	0.65	0.45	0.47	0.57	0.47
S5	0.70	0.55	0.50	0.52	0.47
S6	0.52	0.47	0.45	0.55	0.47
S7	0.50	0.50	0.42	0.50	0.47

Table 3 Matrix of aggregate for all strategies presented

	S1	S2	S3	S4	S5	S6	S7
S1	0	1	1	1	1	1	0
S2	0	0	0	0	0	0	0
S3	0	0	0	0	0	0	0
S4	0	1	0	0	0	1	0
S5	0	1	1	1	0	1	0
S6	0	1	1	0	0	0	0
S7	0	0	0	0	0	0	0

Strategy 1 is identified as the primary measure to prevent ship accidents in ferry and domestic RoRo transportation. Its direct implementation addresses field-level operational issues, including ship stability concerns and operational time constraints within the ferry transport system. The second priority, strategy 5 (enhancement of vessel technical capabilities), emphasizes fleet readiness and requires attention from all operational safety stakeholders. Efforts under this strategy involve reviewing the entire ferry and domestic RoRo fleet, conducting technical condition surveys, and ensuring the precision of sea trial tests. The third recommended action, highlighted by the MADM method, is strategy 4, which focuses on crew training/familiarization pattern changes. This strategy aims to enhance the crew's understanding of the vessel's capabilities, including manoeuvring, stability, and fire containment, through alterations in recruitment and familiarization practices.

Addressing deficiency gaps contributing to the casualties remains crucial, and a comprehensive approach involving multiple strategies is necessary to tackle this chronic issue.

5 Conclusion

Various methods, practical perspectives, and concepts have identified issues in ferry and domestic RoRo ship sinking incidents. Despite the convergence of results, valid data indicates a persistent trend in these accidents. Consistent and effective mitigation efforts prioritizing safety can curb the recurrence of such incidents. While strategy 1 proposes operational field changes, which are crucial for mitigation, pursuing other strategies is essential. These alternatives address different aspects of the ferry and domestic RoRo transportation system, fostering broader workflow changes to enhance maritime safety.

To enhance the study, thoroughly examine each strategy, considering diverse perspectives and criteria. Balancing safety and commercial considerations is crucial, and continuous information gathering for identifying safety issues is vital to maintain dynamism. The anticipated outcomes address current challenges in the domestic ferry and RoRo transportation system.

Acknowledgements The authors wish to thank the Agency of Human Resources Development (BPSDM), Ministry of Transportation, for funding the study, the KNKT for data access, and the respondents for their survey contributions.

References

1. D.L. Ditjen Hubdat, Cetak Biru Transportasi Penyeberangan. Jakarta: Direktorat Lalu Lintas Angkutan Sungai Danau dan Penyeberangan (2010)

2. S. Kristiansen, Maritime transportation: Safety management and risk analysis. Elsevier Butterworth-Heinemann (2005)
3. F. Yusuf, Keselamatan Transportasi (2018). https://setkab.go.id/keselamatan-transportasi/
4. S. Diposaptono, Membangun Poros Maritim Dunia: dalam perspektif tata ruang laut. in *3rd ed. Jakarta: Kementerian Kelautan dan Perikanan*, [Online] (2017)
5. G. Gapasdap, Operasional kapal angkutan penyeberangan yang berkeselamatan, presented at the Acciden review Forum, Merak, Banten (2023)
6. K.V. Dokkum, Ship knowledge: ship design, construction, and operation. in *4th ed. Dokmar maritieme publisher BV* (2016)
7. R. Weisbrod, N. Rahman, Improving ferry safety for lake toba and for the Republic of Indonesia as a whole. WorldBank, Jakarta (2019)
8. J.-U.S.-H. Stephan, P. Assheuer, A.A. Hebbar, Domestic ferry safety in Indonesia: HAZID/ Scoping exercise to identify safety issues pertaining to passenger ships in Indonesia not engaged on international voyages. IMO (2019)
9. I. Komang Hedi Pramana Adiputra, Analysis of domestic ferry safety in Indonesia. World maritime university publisher (2022)
10. IMO, *IMO and ro-ro safety* (International Maritime Organisation, London, 1997)
11. B. Roy, Classement et choix en présence de points de vue multiples. Evue Fr. Inform. Rech. Opérationnelle **2**(8), 57–75 (1968)
12. G.-H. Tzeng, J.-J. Huang, *Multiple attribute decision making: methods and applications* (CRC Press, 2011)
13. OSHA, Using leading indicators to improve safety and health outcomes. U.S. Department of Labor (2019)
14. K. KNKT, Laporan Tahunan Kinerja KNKT. Komite Nasional Keselamatan Transportasi, Jakarta, Annual Report (2022)

Evacuation Simulation with Fire Scenarios on a Passenger Ship

Tezar Pratama and Sunaryo

Abstract Fires on board are a potential risk that ships encounter at sea. Current maritime regulations generally require adherence to prescriptive fire prevention, protection, and extinction. The International Maritime Organization made an advancement by allowing the approval of fire safety designs based on performance-based assessment standards. Meanwhile, passenger awareness of danger indicators and familiarity with escape routes must be investigated to guarantee adequate evacuation time for passengers and crew on board. Fire and evacuation modelling is essential to assess the hazards associated with fire scenarios. This paper simulated the evacuation time from a fire scenario on a 5000 DWT passenger ship with 812 passengers and 30 crew members. The results demonstrate that the evacuation time with the worst fire scenario is 1112 s or 18.6 min. There should be greater reminders to ship passengers to be cautious. This study may serve as a basis for bettering the maritime industry's management of passenger safety.

Keywords Simulation · Fire scenario · Evacuation · Passenger ship

1 Introduction

The challenges surrounding the evacuation of passengers and crews at sea have drawn more attention as a result of the significant loss brought on by serious maritime incidents and accidents due to the rise in the number of high-capacity passenger ships. Compliance with prescriptive regulations set forth by regulatory bodies has

T. Pratama · Sunaryo (✉)
Department of Mechanical Engineering, Universitas Indonesia, Kampus UI Depok, Depok, Indonesia
e-mail: naryo@eng.ui.ac.id

T. Pratama
e-mail: tezar.pratama31@ui.ac.id

Sunaryo
Naval Architecture Study Program, Universitas Indonesia, Kampus UI Depok, Depok, Indonesia

R. Sjarief Widjaja et al. (eds.), *Marine Technology 4*,
SpringerBriefs in Applied Sciences and Technology,
https://doi.org/10.1007/978-3-032-08840-6_10

prevented fire hazards on board ships. These regulations are typically developed in response to well-publicised maritime mishaps, demonstrating a reactive approach to the crucial concern of maritime safety [1]. So far, several guidelines have been released by the International Maritime Organisation (IMO) for conducting full-scale evacuation analyses and certifying large passenger ships in the design stage [2]. A full-scale evacuation simulation is thought to be the most effective technique to show that the situation complies with evacuation standards, but it is typically expensive and challenging to be studied [3]. On the other hand, modelling and simulation tools offer a useful technique to improve ship design concerning evacuation if an accident happens [4].

Drills and evacuation as practice can differ substantially from the evacuation in risky circumstances. Due to ship heeling, the need for passengers and personnel to hold onto handrails to stay balanced on staircases, and perhaps flooded escape routes, flooding can significantly impact evacuees' mobility [3, 4]. The passenger's awareness of danger indicators and familiarity with escape routes must be investigated to guarantee adequate evacuation time for passengers and crew members on board. Aside from the well-known difficulty of evacuation in small shipboard areas, fire effluents also cause evacuees to suffer from extreme heat, toxic fumes, and decreased visibility.

Integrating fire and evacuation models provides a more realistic assessment method by considering the consequences of fire effects on human health. This paper combines the integration of fire scenarios and evacuation modelling to evaluate the safety performance of passenger ships in the event of a major fire scenario. The correspondence by passengers was reported to determine their possible mindset to act during evacuation. It also studied various scenarios of fire load and the location of the initial fire point to consider the worst situation and develop an evaluation protocol in ship safety management.

2 Methodology

This work simulated the evacuation of passengers on a 5000 DWT passenger ship with four scenarios. Two scenarios are assumed without a fire occurrence to determine how long it would take to evacuate every passenger to a safe place. These scenarios are then compared with two other scenarios considered if a fire breaks out and blocks one of the room doors. The ship's capacity is 812 passengers and 30 crew members. Then, the ship's general arrangement (GA) was investigated to design fire scenarios and evacuation modelling based on the layout of ship compartments (see Fig. 1).

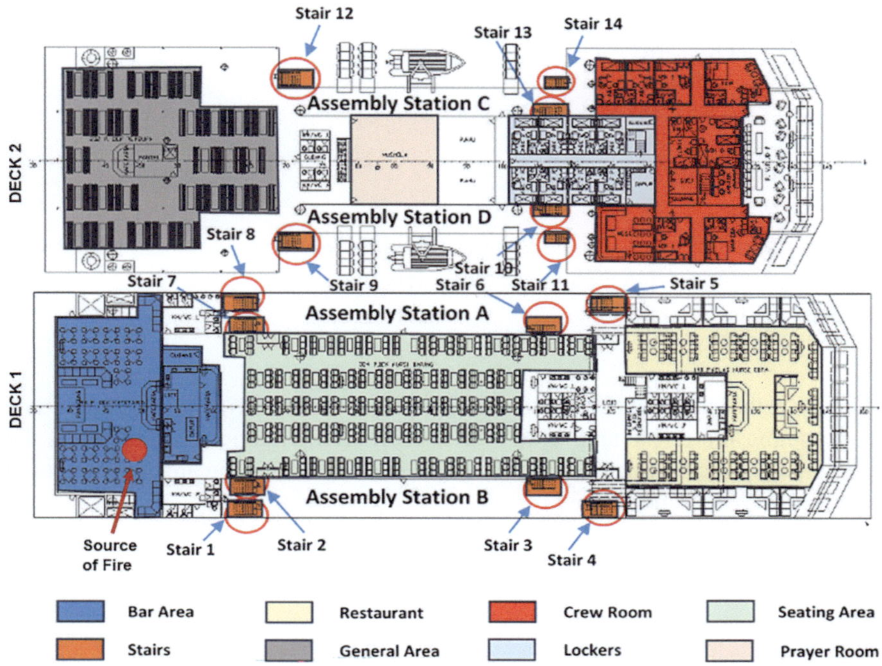

Fig. 1 The passenger ship layout with highlighted fire-starting locations and assembly stations

2.1 *Fire Scenario*

Deck 1 of the passenger compartment on the starboard side of the ship was chosen as the fire's starting point (see Fig. 1). The location being chosen is the bar area, which is a spot where people on board frequently visit. The bar area was designed for a total maximum capacity of 72 people. In this simulation, four scenarios are created. Scenario A is a condition where there is no fire, so that all doors can be used as evacuation routes, and the total number of people in the bar area is 72 people, while scenario B has the same condition but assumed the number of people is twice. Scenario C is to simulate the same number of people as A, but there is a fire, so that one of the doors in the bar area was closed, so that 72 people in the room could only access one door for the evacuation route, and scenario D was a fire with a total of 144 people in the bar area. To assume a fire event requires knowing the energy release that occurs and its duration, where mathematically, it can be calculated by:

$$\dot{Q} = \dot{m}\Delta H_{\text{eff}} \tag{1}$$

where $\dot{m}$ is given in kilograms per second or grams per second, and ΔH_{eff} is the effective heat of combustion, often given in kilojoules per kilogram or kilojoules per gram. The resulting energy release rate will then be in kilojoules per second, which is the same as kilowatts.

2.2 Evacuation Modelling

The time for evacuation must be less than the time for the environment in that ship to become life-threatening, inclusive of a safety margin:

$$t_{\mathrm{ev}} + t_s < t_{\mathrm{lt}} \tag{2}$$

where t_{ev} is the calculated evacuation time measured from the ignition, t_{lt} is the time for becoming life-threatening again measured from the ignition, and then t_s is the safety margin. Evacuation time and time become life-threatening are both measured from the time of ignition, given by:

$$t_{\mathrm{ev}} = t_d + t_a + t_o + t_i + t_t + t_q \tag{3}$$

where t_d is the time from ignition until detection of the fire (by the occupant or by an automatic detection system), t_a is the time from detection until the alarm is sounded, t_o is the time from the alarm until the time of occupants make a decision to respond, t_i is the time for occupants to investigate the fire, collect belongings, or fight the fire, t_t is the travel time, being the actual time required to transverse the escape route until a place of safety is reached, including way-finding, and t_q is the queuing time at doorways or other obstructions. The term t_d can be assumed to be 0.75 min [5], and the term should be estimated from knowledge of the alarm, which was assumed to be 0.2 min. The terms t_o an t_i are more difficult to calculate because it is human behaviour analysis. This work also tries to analyse correspondents. Travel time (t_t) and t_q were calculated using evacuation modelling.

Evacuation modelling can be performed using the information on walking speeds found in IMO MSC. Circ 1533. This work used the software Pathfinder 2023 to simulate the evacuation. This simulation was used to learn about the behaviour of passengers, such as when they wait for a specific reaction time for a predetermined amount of time, and when the agent goes between waypoints, rooms, and exits.

Then, the passenger categories are assumed based on their age. Based on other studies [6], it was reported that in Indonesia, the distribution of ship passengers is as follows: women under 30 years old make up 12% of the total, women between 30 and 50 years old account for 21%, women over 50 years old account for 5%, and there are no female passengers with disabilities. Meanwhile, men under 30 years old make up as much as 19% of the total, men between 30 and 50 years old account for 31%, men over 50 years old account for 12%, and there are no male passengers with disabilities.

2.3 Investigation of Passenger Behaviour

To analyse passenger behaviour in the event of a ship fire, secondary data were gathered for this study by having 60 passengers serve as correspondents and fill out questionnaires. Each age category was represented by 10 correspondents. After the passenger ship safety management gave the safety induction communication, responders were questioned about their familiarity with evacuation routes and self-rescue approaches. The simulation and analysis both consider the findings of the inquiry of the correspondent's responses. In emergency maritime scenarios, it is crucial to make the evacuation effective as much as possible to reduce evacuation time and fatalities [7]. The key benefit of considering passenger behaviour is the ability to create an efficient emergency evacuation system to maintain safety standards.

3 Results and Discussion

3.1 Passenger Compliance Behaviour

To track the evacuation process more efficiently, it can be crucial to comprehend the compliance psychology of the passengers. Additionally, it complies with the advanced evacuation analysis recommendations that consider passengers' emotions. According to Chu et al. [8], compliance behaviour refers to how people behave by what is expected of them, given rules, guidelines, and the opinions of others. The degree of compliance can be affected by both environmental and human factors [9]. In another work, Jorgensen and May [10] conducted an empirical study on passengers' attitudes and actions about disobeying instructions and how this may affect how long people assemble in an emergency aboard a ship. Understanding this human characteristic, considering the compliance class of evacuees may thus represent another future line of inquiry in this field. The percentage of responders who correctly comprehended the passenger safety instructions that were presented during the safety induction before the ship departed is shown in Fig. 2. A score of 1 to 5 indicates that passengers understand the evacuation procedures and know the evacuation route, 1 indicates that passengers do not understand, and 5 indicates that they really understand.

3.2 Passenger Risk Perception

The crucial decision for passengers to move after recognising the initial indications occurs during the response time. The passengers' risk perception (RP) is the key factor influencing this choice [11]. Passengers' subjective assessments of the likelihood of unfavourable outcomes, such as death and ambiguity in the evacuation procedure, are

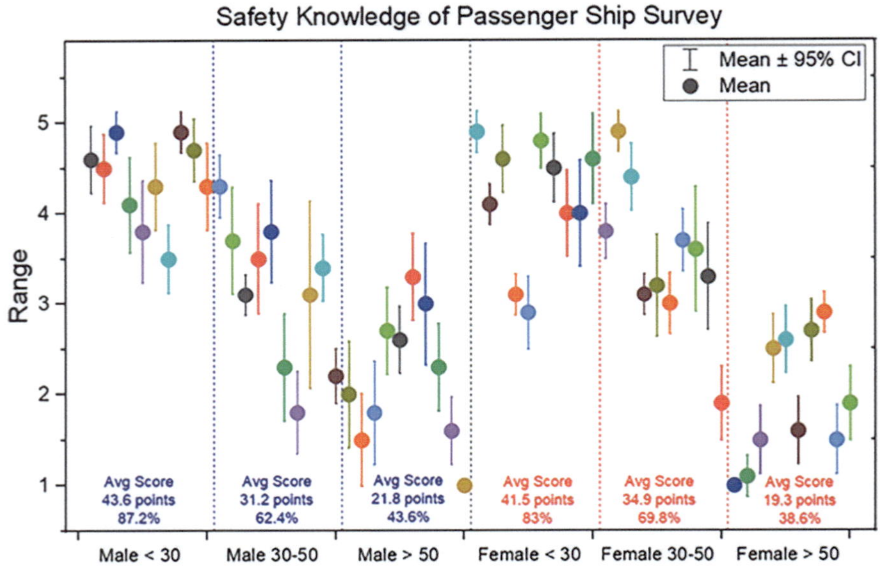

Fig. 2 Safety knowledge of passenger ship

the root causes of RP. The RP factor should be considered since it helps us understand how people think. It may also help us reduce the overall evacuation time. From the results of the correspondence, it was found that passengers took an average of 77.9 s or 1.3 min to realise the position and number of evacuation routes on the ship (see Fig. 3). Most of them required more time. This could be because passengers were unfamiliar with the introduction of evacuation routes and did not pay attention when the safety induction was presented, as well as the large shape of the ship, allowing passengers to experience disorientation while inside the ship.

3.3 Evacuation Time by Simulation

Based on the simulation results, the travel time of a total of 812 passengers with 30 crews is shown in Fig. 4. The simulation results show that scenario A takes 1027.53 s to evacuate 843 people, compared to scenario C, with fire conditions requiring almost the same total time. However, if a fire had happened in the bar area as simulated in scenario C, it took 103 s, while scenario A only took 69 s to evacuate 72 passengers. Then, in the comparison between scenario B and scenario D, the scenario without a fire in the bar area requires a longer evacuation, namely 348 s. At the same time, scenario D only requires 198 s to evacuate 144 people from the bar area if a fire occurs. This is because the exit door from the bar area is very close to the other escape doors, so if people try to use the same door, it will cause a longer queuing time. This can be seen in more detail in Fig. 4, where passengers were able to leave

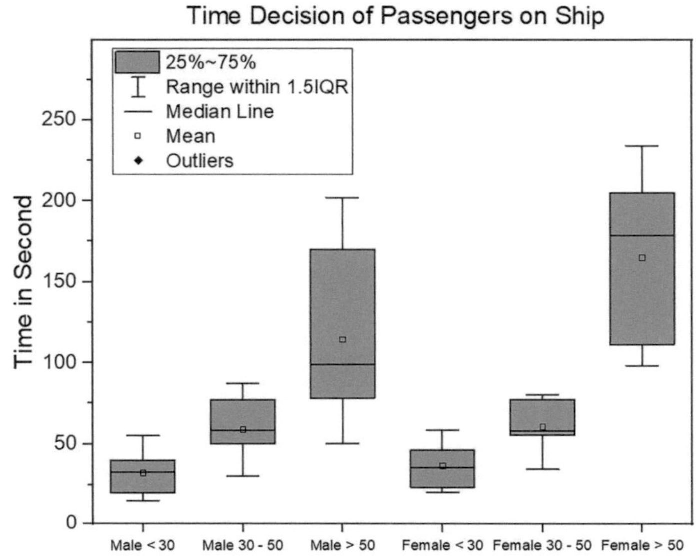

Fig. 3 Indication risk time of passenger ship

the bar area in around 152 s, but there was a long queuing time between the next doors on the evacuation route. In contrast, if one of the doors is closed due to a fire, this will actually make the occupants walk around in a direction so that there is time to get to the exit and minimise queuing time. However, scenario C requires an evacuation time of 1030.53 s, while scenario D is 1035.03 s.

The computation of the overall evacuation time can be used to update the full evacuation duration in a real-time emergency response. It can specifically assist crew and passengers in managing the remaining time [12] based on the evacuation routes that are accessible. However, it does have certain shortcomings. It neglects to consider passengers as aware agents, for example, in a real situation. The diversity of passenger characteristics, such as their level of compliance, may expressly have an impact on the overall evacuation time.

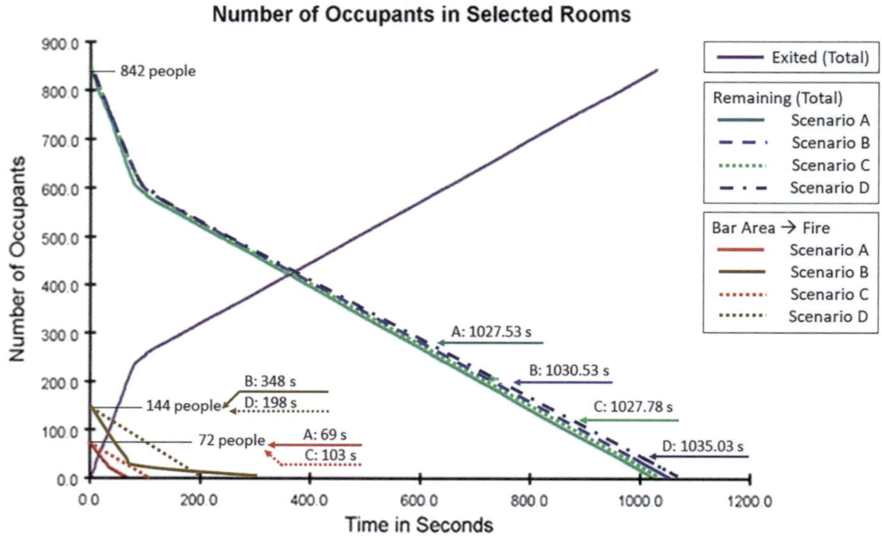

Fig. 4 Number of occupants in selected rooms (bar area), which were scenarios **a** and **b** without a fire, scenarios **c** and **d** with a fire

4 Conclusion

The evacuation time from a fire scenario on a passenger ship with total 842 people was studied by simulation to understand how passengers would behave in the event of a fire and estimate the necessary evacuation time, 60 persons were surveyed. The identification results demonstrate that it takes an average of 77.9 s for passengers to select to evacuate and comprehend the closest evacuation route from their location, which may have an impact on the evacuation procedure. If calculated in the simulation, the total evacuation time with the worst fire scenario is 1112.93 s or 18.55 min. There should be greater reminders to ship passengers to be cautious. This study may serve as a basis for bettering the maritime industry's management of passenger safety.

References

1. C. Azzi, A. Pennycott, G. Mermiris, D. Vassalos, Evacuation simulation of shipboard fire scenarios. in *Fire and evacuation modeling technical conference, baltimore, Maryland, 15–16 August 2011*
2. International maritime organization: Guidelines for evacuation analysis for new and existing passenger ships. https://www.imorules.com (2007). Accessed 11 May 2023
3. W.L. Wang, S.B. Liu, S.M. Lo, L.J. Gao, Passenger ship evacuation simulation and validation by experimental data sets. Procedia Eng. **71**, 427–432 (2014)
4. D. Vassalos, C. Azzi, A. Pennycott, Crisis management onboard passenger ships. in *International conference on human performance at sea, Glasgow, Scotland, 16–18 June 2010*

5. H.E. Nelson, FPETOOL: Fire protection engineering tools for hazard estimation, NIST Internal Report. https://www.nist.gov/publications/fpetoolfire-protection-tools-hazard-estimation. Accessed 11 May 2023
6. M.B. Zaman, N. Siswantoro, M. Zulfikar, Evacuation analysis of ship passenger using agent based modelling simulation. Earth Environ. Sci. **698**(1), 012017. IOP Publishing (2021)
7. O. Finiti, Understanding and predicting human behaviour in maritime emergencies. Dissertation, University of Huddersfield (2021)
8. C.W. Chu, H.A. Lu, C.Z. Pan, Emergency evacuation route for the passenger ship. J. Mar. Sci. Technol. **21**, 515–521 (2013)
9. K. Hamad, A. Faghri, R. Nanda, A behavioral component analysis of route guidance systems using neural networks. Comput. Civ. Infrastruct. Eng. **18**, 440–453 (2003)
10. H.D. Jorgensen, M. May, Human factors management of passenger ship evacuation. in *RINA Conference, London, 2–3 Oct 2002*
11. M.T. Kinateder, E.D. Kuligowski, P.A. Reneke, R.D. Peacock, Risk perception in fire evacuation behavior revisited: definitions, related concepts, and empirical evidence. Fire Sci. Rev. **4** (2015)
12. C.S. Lin, M.E. Wu, A study of evaluating an evacuation time. Adv. Mech. Eng. (2018). https://doi.org/10.1177/168781401877

Analyzing the Types of Potential Return Cargo for Livestock Carriers to Stimulate Economic Movement in Livestock-Producing Regions

Kusuma Dhipanusa, Tri Achmadi, and Bayu Saputra Syah

Abstract The Tol Laut is a priority program of the Indonesian government to connect one region with other regions separated by the ocean using sea transport to ensure the implementation of a sustainable logistics flow, so that it is expected to reduce price disparities between one region and another in Indonesia. Nusa Tenggara Timur and Nusa Tenggara Barat are the suppliers of fresh beef to several provinces in Java and Kalimantan. However, transporting live cattle using cargo ships has the potential to cause cattle weight loss of up to 15–20%. In 2015, the Ministry of Transportation operated a livestock carrier that applied animal welfare principles so that cattle's weight loss decreased to only 9%. In its operation, the livestock carrier showed an increase in the number of livestock transported, but the ship always returned empty to livestock supply areas. This paper will then discuss potential solutions that can be presented to increase the productivity of livestock carriers using KM. Camara Nusantara 1 as an example. The results of this discussion are the types of return cargo that can be loaded on livestock carriers that can balance the cost of logistics distribution while encouraging economic movement in livestock supply areas.

Keywords Tol laut · Livestock carrier · Return cargoes · Indonesia

K. Dhipanusa (✉) · T. Achmadi · B. S. Syah
Institut Teknologi Sepuluh Nopember, Kampus ITS, Sukolilo, Surabaya, Kota SBY, Jawa Timur, Indonesia
e-mail: k_dhipa@yahoo.com

T. Achmadi
e-mail: triachmadi@seatrans.its.ac.id

B. S. Syah
e-mail: bsaputrasyah@gmail.com

R. Sjarief Widjaja et al. (eds.), *Marine Technology 4*,
SpringerBriefs in Applied Sciences and Technology,
https://doi.org/10.1007/978-3-032-08840-6_11

1 Introduction

Indonesia is the largest archipelagic country with 17,001 islands [1]. The area of Indonesia is divided into a land area of 1.91 million square kilometers and a water area of 6.32 million square kilometers. A shift in government perspective sees the water as a connector, not a separator. To realize this, they are developing the maritime sector, including the Tol Laut project.

The principle of the Tol Laut program is to build a sea transportation system that connects all parts of Indonesia from Sabang to Merauke. It consists of six shipping routes that are barrier-free, connecting ports throughout Indonesia so that they connect between islands [2]. Tol Laut was originally introduced as a concept of transporting marine logistics connecting ports in Indonesia with the main objective of reducing price disparities that occur between Java and other islands [3].

The livestock carrier that was built is a ship specifically designed for the transportation of livestock, especially cattle, by taking into account the principles of animal welfare [4]. The construction of livestock carriers is now crucial for animal welfare, particularly following research on cattle transportation by ship. Findings revealed a weight loss of up to 17.22% during a journey from Lombok to West Kalimantan [5]. For the transport of cattle from Nusa Tenggara Timur (NTT) to Jakarta, the weight loss during transportation reduced from 15 to 20% on cargo ships to only 9% with livestock [6].

The number of fleets and routes of livestock carriers has continued to increase since 2015 until now (see Fig. 1) with good performance where the number of livestock loads has shown an increase from year to year (see Fig. 2). Since 2021, the Tol Laut program has begun to focus more on return cargo to balance logistics distribution costs and stimulate the economy in the 3 T (frontier, remote, and underdeveloped) areas. [7].

For return cargoes that can be transported on livestock carriers are items that are not easily contaminated with the smell of livestock cages and also do not risk damaging the cages on the ship [8].

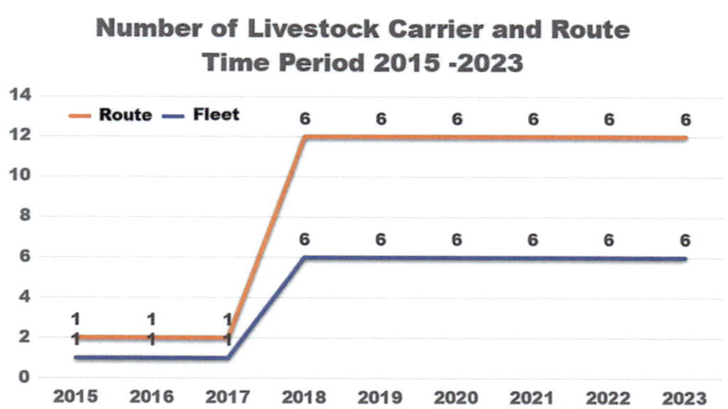

Fig. 1 Number of fleet and routes

Fig. 2 Realization of livestock load

2 Methodology

In this study, data collection uses secondary data derived from several sources in the form of publication results and reports of coordination meetings of related agencies. Focusing on KM. Camara Nusantara I as the object of research. The research method used is a descriptive and qualitative design. Starting from data collection, analysis, and finally formulated into a conclusion. Data analysis is carried out by looking at trends in community needs in livestock-producing areas (NTT), and then analyzed by looking at the characteristics of products that have the potential to become return cargoes.

3 Results and Discussion

The economic structure in NTT is dominated by the agriculture, forestry, and fisheries sectors. The demand for rice in NTT increases significantly along with population growth (see Fig. 3: NTT population [16] Fig. 3). However, this was not in line with rice production in NTT, which showed a decline from 2018 (see Fig. 4).

This has contributed to higher rice prices due to scarcity in the market. The non-ideal rice yield in NTT is also a factor that exacerbates the rice crisis in NTT [9]. The government responded to this need by distributing imported rice from Vietnam [10].

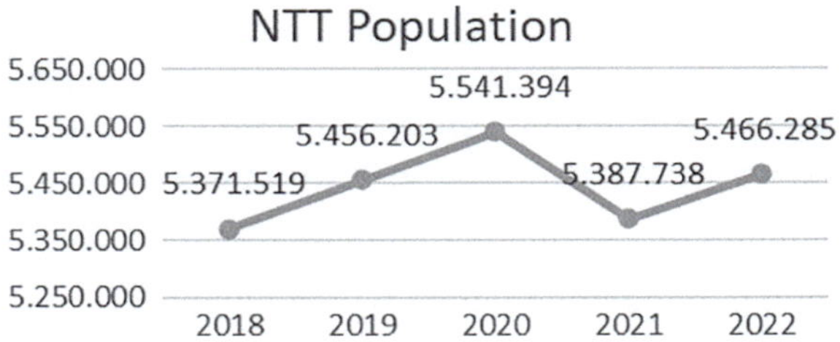

Fig. 3 NTT population [16]

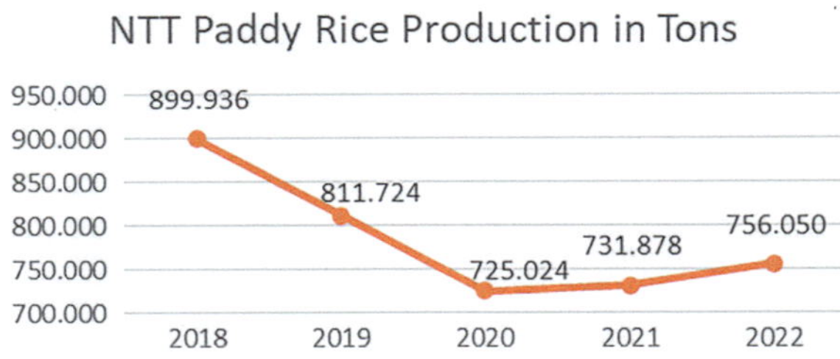

Fig. 4 Paddy rice production [17]

Infrastructure is one of the factors that boost regional productivity. Road, electricity, and water infrastructure together have a significant influence on economic productivity in NTT province [11].

3.1 KM Camara Nusantara I

MV. Camara Nusantara 1 is a specialized livestock carrier operated by PT PELNI. It has a size of 1587 gross tonnage with a maximum carrying capacity of 550 cattle in one transport. This ship has a route Kupang-Waingapu-Tanjung Priok-Patimban-Kupang with a total distance of 2256 nautical miles [12], this ship also has a hatch with a capacity of 150 tons.

In the operation of MV. Camara Nusantara I there are several notes raised by PT PELNI as the operator, among others: (1) the cost of ship maintenance is very large because spare parts and ship equipment are still imported so it depends on the rupiah exchange rate, (2) gross and net profit margins are still below 10% so that optimization and efficiency are needed, (3) subsidies from the government still reach 80% [13].

3.2 Return Cargoes

Return cargo goods that have the potential to be transported as return cargo by livestock carriers (MV Camara Nusantara I) must be commodities that are needed by the people of NTT and are able to encourage economic growth there. Commodities that will be used as return cargo must be goods produced around the unloading port area, namely Tanjung Priok and Patimban, so that the initial price is cheaper than similar goods produced in areas closer to NTT, such as Surabaya and Makassar, so that the price of commodities when they arrive in NTT is still cheaper.

Related to the three strategic issues developing in NTT, the need to improve the economy in NTT includes: (1) staple food (rice); (2) fertilizer and complete animal feed; (3) construction raw materials.

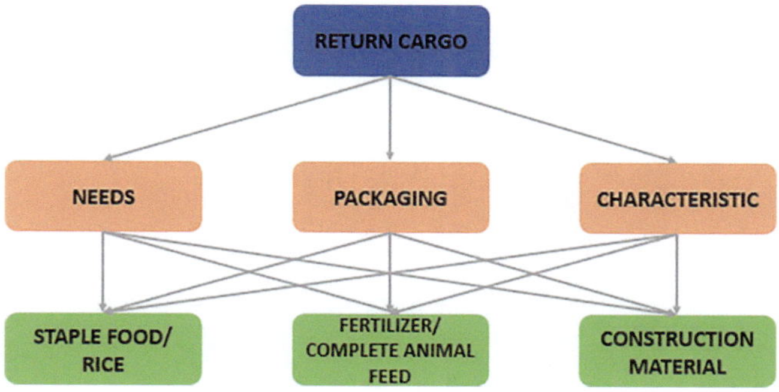

Fig. 5 Analytical hierarchy process (AHP)

Using the analytical hierarchy process (AHP) method (see Fig. 5) to measure the weight of each choice based on criteria calculated using Excel (see Table 1). The results of calculations using AHP are used to rank the most suitable return cargo to be transported by livestock carriers (see Table 2).

Staple foods such as rice are not ideal for return cargo on livestock carriers due to the potential for spoilage due to the smell of the barn, unless the rice is packed in airtight packaging (rather than being loaded in sacks or bulk).

Fertilizer becomes a necessity for NTT residents. It has characteristics that fall into the class of smelly cargo and dirty cargo. So, this cargo has great potential to be used as return cargo because it can be loaded in the hold space or can even be maximized by using the cage space.

Furthermore, complete animal feed also has the potential to be a return cargo for livestock carriers. This is because NTT is one of the priority areas for beef cattle production centers for limited resource efficiency in order to continue to produce highly competitive beef cattle [14].

Construction materials are one of the most important commodities in infrastructure development in NTT. Construction activities on roads and buildings in NTT have a more significant influence on economic growth than other construction activities [15]. These commodities tend to be large in density, so the volume of goods that can be transported will tend to be small, with the available cargo hold capacity of only 150 tons.

Table 1 Criteria weight

Criterion	Needs level	Packaging	Characteristics	Eigen			Average
Need level	1.0	5.0	1.0	0.4545455	0.3846154	0.4666667	0.04352758
Packaging	0.2	1.0	0.1	0.0909091	0.0769231	0.0666667	0.0781663
Characteristics	1.0	7.0	1.0	0.4545455	0.5384615	0.4666667	0.4865579
	2.2	13.0	2.1	1.0	1.0	1.0	

Table 2 Return cargoes ranking

Return cargo		Ranking
Staple food	0.3127221	2
Fertilizer/animal feed	0.4158349	1
Construction material	0.2714430	3

4 Conclusion

The return cargo that can be transported by livestock carriers must be cargo that is needed by the people of NTT to support the economy. Maintaining the quality of return cargo to NTT requires appropriate cargo characteristics and loading methods to prevent damage.

From the discussion in this study, fertilizer and complete feed are the most suitable commodities to be backloaded. This is inseparable from the characteristics of these two commodities, which tend to be resistant to the aroma of the cage.

The author suggests packing these two commodities using sacks so that they can utilize the hatch space to the maximum, and in extreme conditions are still able to be stacked to fill empty cage space. Transporting rice as return cargo is feasible, but it requires careful consideration of packaging and loading procedures to prevent damage. Further analysis is needed for the selection of rice as return cargo.

The author's hope for this study is that it can be used to increase the interest of entrepreneurs to utilize the empty space of livestock carriers and boost the economy of NTT residents, and help the government spend on livestock carrier operating costs.

Acknowledgements The authors gratefully acknowledge the Directorate of Sea Transportation for providing data and information to the author. Thanks to Institut Teknologi Sepuluh November (ITS) for the guidance from the Lecturers in compiling this article.

References

1. Annur, Cindy Mutia, Ada 17.001 Pulau di Indonesia pada 2022, Ini Provinsi dengan Pulau Terbanyak, 28 April 2023. [Online]. Available: https://databoks.katadata.co.id/demografi/statistik/3b74360c445ca46/ada-17001-pulau-di-indonesia-pada-2022-ini-provinsi-dengan-pulau-terbanyak
2. Humas, Presiden Jokowi Jelaskan Konsep Sebenarnya Tol Laut, 20 Februari 2017. [Online]. Available: https://setkab.go.id/presiden-jokowi-jelaskan-konsep-sebenarnya-tol-laut/
3. Divisi Informasi, Rute Tol Laut Idealnya Hanya Melayari Jalur Perintis, 14 September 2017. [Online]. Available: https://supplychainindonesia.com/rute-tol-laut-idealnya-hanya-melayari-jalur-perintis/
4. Admin Portal, Kemenhub Luncurkan Kapal Khusus Ternak Pendukung Program Ketahanan Pangan, 22 November 2017. [Online]. Available: https://hubla.dephub.go.id/home/post/read/5446/kemenhub-luncurkan-kapal-khusus-ternak-pendukung-program-ketahanan-pangan#:~:text=%22Dengan%20adanya%20kapal%20pengangkut%20ternak,dapat%20terpenuhi%2C%22%20lanjut%20Wisnu

5. A. Anton, L.M. Kasip, L. Wirapribadi, S.N. Depamede, A.S. Asih, Perubahan Status Fisiologis dan Bobot Badan Sapi Bali Bibit yang Diantarpulaukan dari Pulau Lombok ke Kalimantan Barat. J. Ilmu Dan Teknol. Peternak. Indones. **2**, 86–95 (2016)
6. Kapal Ternak Tekan Risiko Penyusutan Bobot Sapi Hingga 9, 8 Februari 2019. [Online]. Available: https://pilarpertanian.com/kapal-ternak-tekan-risiko-penyusutan-bobot-sapi-hingga-9
7. Biro Komunikasi dan Informasi Publik, Program Tol Laut Fokus Angkutan Balik, 2 July 2021. [Online]. Available: https://kemenhub.go.id/post/read/program-tol-laut-fokus-angkutan-balik
8. Divisi Informasi, Angkutan Laut Bersubsidi: Pebisnis Bisa Isi Muatan Balik Kapal Ternak, 07 Januari 2019. [Online]. Available: https://supplychainindonesia.com/angkutan-laut-bersubsidi-pebisnis-bisa-isi-muatan-balik-kapal-ternak/
9. F.P. Herin, Krisis Beras, NTT Sedang Tidak Baik-baik, 20 March 2023. [Online]. Available: https://www.kompas.id/baca/nusantara/2023/03/17/krisis-beras-ntt-sedang-tidak-baik-baik
10. A. Dhema, Sediakan 10.000 Ton, Bulog Jamin Ketersediaan Beras di NTT Hingga Semester I 2023, 26 January 2023. [Online]. Available: https://kupang.tribunnews.com/2023/01/26/sediakan-10000-ton-bulog-jamin-ketersediaan-beras-di-ntt-hingga-semester-i-2023. Accessed 10 September 2023
11. L.C. Gunita, M.Y. Luruk, N. Tameno, Pengaruh Infrastruktur Terhadap Produktivitas Ekonomi di Provinsi NTT, J. Ekon. **4** (2019)
12. PT Pelayaran Nasional Indonesia, Laporan Tahunan 2018 Annual Report, PT Pelayaran Nasional Indonesia, Jakarta (2018)
13. D. Yuwono, S. Untung, F. Tantri, PT Pelni's Operational Performance as a Livestock Ship Operator, J. Manaj. Transp. Dan Logist. (JMTRANSLOG). **6** (2019)
14. Y. Yusdja, N. Ilham, Tinjauan Kebijakan Pengembangan Agribisnis Sapi Potong, Anal. Kebijak. Pertan. **2** (2004)
15. M. Luter, I. Indrocahyo, I.L.V. Resti, Pengaruh Belanja Infrastruktur terhadap Pertumbuhan. J. Ekon. Pembang. **17**, 38–48 (2019)
16. BPS, Jumlah Penduduk Menurut Jenis Kelamin (Jiwa), [Online]. Available: https://ntt.bps.go.id/indicator/12/556/2/jumlah-penduduk-menurut-jenis-kelamin.html. Accessed 10 September 2023
17. BPS, Produksi Padi menurut Kabupaten/Kota (Ton), [Online]. Available: https://ntt.bps.go.id/indicator/53/929/2/produksi-padi-menurut-kabupaten-kota.html. Accessed 10 September 2023

Measuring the Role of Subsidized Shipping in Improving Economic and Social Welfare: A Case Study in the Tanimbar Islands

Pratiwi Wuryaningrum, Firmanto Hadi, Irwan Tri Yunianto, Siti Dwi Lazuardi, Dika Virginia Devintasari, and Masfufatuzzama Rida

Abstract Indonesia, as an archipelagic country, has an uneven population distribution and limited economic activity. Regarding this fact, sea transportation has a significant role in improving welfare, especially for those who live in particular areas, which are in Indonesia called 3TP areas (remote, outermost, and border areas). To spur the development of the 3TP areas, the government has implemented a policy of providing subsidized shipping (non-commercial shipping), which is called "Perintis" shipping. This service connects remote areas in order to improve accessibility. Subsidized shipping has the characteristic of 'Transport Promote the Trade,' so it cannot be judged as profit-oriented commercial transportation in determining its success. The aim of this paper is to assess the benefits of subsidized shipping based on economic factors and social welfare. This research is conducted in a case study of the Tanimbar Islands district, which is a district that is included in all the criteria for the 3TP area, and using the analytical hierarchy process method. The results show that the Tanimbar islands district is included in the category of underdeveloped and outer/border region, which has the percentage of each criterion as follows: politics

P. Wuryaningrum (✉) · F. Hadi · I. T. Yunianto · S. D. Lazuardi · D. V. Devintasari
Department of Marine Transportation Engineering, Institut Teknologi Sepuluh Nopember Surabaya, Surabaya, Indonesia
e-mail: pratiwi@its.ac.id

F. Hadi
e-mail: firmanto@its.ac.id

I. T. Yunianto
e-mail: irwantri.y@its.ac.id

S. D. Lazuardi
e-mail: dwilaz@seatrans.its.ac.id

D. V. Devintasari
e-mail: dika.virginia@its.ac.id

M. Rida
Department of Industrial and Data Engineering, Pukyong National University, Busan, South Korea
e-mail: rida1403@pknu.ac.kr

R. Sjarief Widjaja et al. (eds.), *Marine Technology 4*,
SpringerBriefs in Applied Sciences and Technology,
https://doi.org/10.1007/978-3-032-08840-6_12

is 26%, economics is 14%, social-cultural is 8%, and security and defense is 52%. From the results of the benefit calculation, the Tanimbar Islands district is a region with a high benefit value, namely 0.0188.

Keywords Subsidized shipping · Remote areas · Social welfare

1 Introduction

Indonesia's geographical condition as an archipelago with unequal population distribution has resulted in some regions having lower population density than others. Consequently, these isolated areas, called remote areas, have low accessibility and minimal economic activity. Thus, their development lags behind that of regions with better accessibility. This poses significant challenges to sea transportation and logistics in the archipelago [1]. Then, to connect one region to another, transportation facilities are needed [2].

Residents in remote areas have low purchasing power, making it economically unviable for businesses to offer transportation services. The government provides Perintis Shipping transportation to promote the development of these remote regions [3]. According to [4], it is important in facilitating economic growth. Furthermore, another study shows that maritime container transport is an important factor in stimulating economic growth [5]. Additionally, researchers such as [6] have stated that improvements in transportation and distribution can bring about significant changes in production areas or vice versa.

Perintis Shipping is a subsidized service that connects remote areas in Indonesia to improve community accessibility and spur economic development. The region has unique characteristics, including dangerous shipping conditions in the east that require large ships for transportation, but with low cargo demand, making it unprofitable for commercial shipping firms. Therefore, government involvement is crucial.

Based on the problem statement, the proposed research aims to determine the indicators for assessing and measuring the role of Perintis Shipping services in the district of Tanimbar Island.

2 Methodology

There are several steps in the research methodology. The flowchart of this research can be seen in Fig. 1. The research employs the analytical hierarchy process (AHP) method.

It will start the analysis by conducting a 3TP regional development assessment that will focus on economic growth and the human development index. This assessment will also take into account the current state of pioneer transport, including its

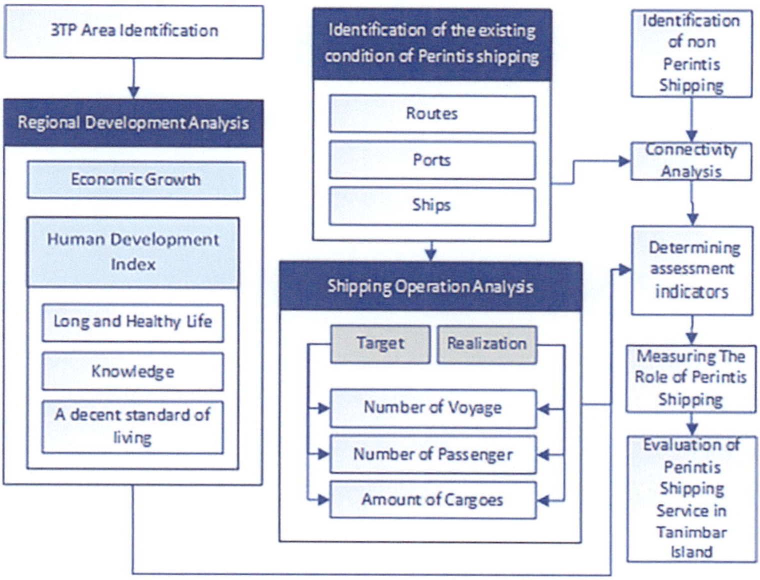

Fig. 1 Research methodology

routes, ports, and ship operations. Then, examining the connectivity of areas that are serviced by pioneer transport as well. Based on our findings, we will analyze the shipping operations, including the number of voyages, passengers, and cargo carried by the ships. Our next step is to determine assessment indicators by considering the development of the 3TP area, analyzing ship operations, and examining regional connectivity. Once we have established these assessment indicators, we use them to evaluate the role of pioneer transportation in the area being studied. Based on our measurements, we are able to assess whether or not pioneer transportation should continue.

3 Results and Discussion

3.1 Identification of Underdeveloped, Outer, and Border Areas

Underdeveloped areas, as defined by Regulation of President 63/2020, refer to districts with less developed societies and territories compared to other regions on a national scale. The location of the 3TP Area and areas served by Perintis Shipping (marked green) are shown in Fig. 2.

The outer area consists of small outer islands (PPKT) with an area of < 2000 km^2. It has geographic coordinate base points that connect the archipelagic sea baselines by

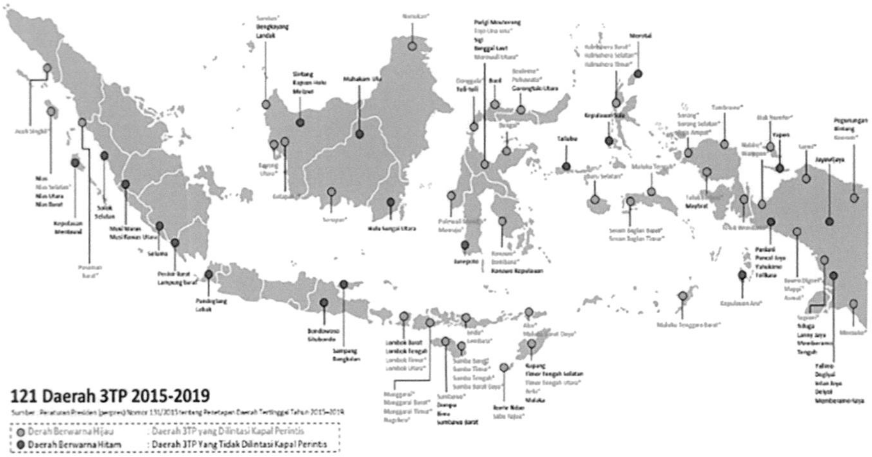

Fig. 2 Location of the 3TP area and areas served by Perintis Shipping (marked green)

international and national law. PPKT is also designated as a KSNT (certain national strategic area), namely an area related to state sovereignty, environmental control, and/or a world heritage site, the development of which is prioritized for national interests (Directorate General of Marine Spatial Management, n.d.).

According to the Law of the Republic of Indonesia, Number 43 of 2008, border areas are parts of the country's territory located on the inside along the borders of Indonesia's territory with other countries. In terms of state territorial boundaries on land, the border areas are in sub-districts.

3.2 Identification of Perintis Shipping

Route, Ship, and Ports

The mode of transportation for Perintis Shipping was the Perintis ship. The Perintis ship is a multipurpose ship that carries passengers and goods. These Perintis ships have several types and capacities, including 200 DWT, 500 DWT, 750 DWT, 1200 GT, and 2000 GT. The Perintis Shipping route served marine transportation throughout Indonesia. There are 117 routes, and there are 521 transit ports and 42 home ports in 2022 [7].

3.3 Perintis Shipping Services in the District of Tanimbar Island

Tanimbar Islands Regency is one of the districts in Maluku province and is a division of the Southeast Maluku regency area, as shown in Fig. 3. The Tanimbar Islands Regency area is included in underdeveloped areas and/or remote areas, border and outer areas (3TP). Apart from that, the Tanimbar Islands Regency area is also an archipelago, so this area is used as a case study in this research. There are two ports visited by the pioneer ship, namely Saumlaki port and Tutu Kembong port. There are five Perintis routes serving this district, namely routes R-24, R-27, R-70, R-72, and R-75. This is the location of the Tanimbar Islands. Statistical data of the district of Tanimbar island can be found in Table 1.

Fig. 3 Distric of Tanimbar island

Table 1 Statistical data of district of Tanimbar island

Parameter	Year		
	2017	2018	2019
GDRB (Million-Rp)	1.500,79	1.590,72	1.685,81
IPM	61,64	62,39	62,86
Population	111.825,00	125.736,00	127.365,00
Number of Voyage	130	222	204
Number of passenger (Pax)	41.218	59.411	79.083
Amount of cargo (Ton)	10.593,00	8.765,23	9.631,70

Table 2 Assessment indicators

Sector Parameter	Description of perintis shipping role	Assessment indicators
Political	As a means of supporting the implementation of government administration throughout the country; As a connecting network and means of unifying the Republic of Indonesia in supporting the realization of the archipelago's insight	Perintis shipping connectivity in the 3TP area
Economics	Facilities to support the movement of people and/or goods; Means of stimulating regional economic growth (ship promotes the trade)	Correlation of perintis shipping with GRDP
Social-Cultural	A means of increasing mobility and social and cultural interaction between citizens of the nation	Correlation of perintis shipping with HDI
Security and Defense	National merchant fleet as a component of national defense; The national merchant fleet as a supporter of national defense at sea when the country is in a state of danger can be mobilized	Identify 3TP areas that are included in the Islands Area Category

3.4 Analysis of Determining Assessment Indicators of the Perintis Role

Assessment Indicators

Determination of the calculation of the Perintis ship services, based on the role of Perintis Shipping in the fields of political, economic, social-cultural, and security and defense, which are described in Table 2.

Calculation of the Weight of Assessment Indicators

Assessment of indicator weights is carried out using the AHP method by assessing each regional category. The regional category is divided into three, namely the category of underdeveloped areas, the category of border areas, and the category of underdeveloped areas that are included in border areas.

Indicators Weight Calculation

The weighting results for the three regional categories are shown in Table 3 with a sample of underdeveloped and border areas. From the results of the weighting for each regional category, the results of the recapitulation of the weights of each indicator for the three regions are obtained as shown in Table 4. For the underdeveloped areas, the dominant parameter is the economic sector, accounting for 43%. Meanwhile, for

Table 3 Indicators weight calculation sample on underdeveloped and border area

Underdeveloped and border area				
	Political	Economic	Social-Cultural	Security and defense
Political	1	3	4	1/4
Economic	1/3	1	2	1/3
Social-cultural	1/4	1/2	1	1/5
Security and defense	4	3	5	1

Table 4 Result of criteria for each area category

Area category		**Criteria**	
Underdeveloped area	100%	Political	9%
		Economic	**43%**
		Social-cultural	39%
		Security and defense	9%
Border area	100%	Political	13%
		Economic	10%
		Social-cultural	21%
		Security and defense	**56%**
Underdeveloped and border area	100%	Political	26%
		Economic	14%
		Social-cultural	8%
		Security and defense	**52%**

the border areas, the dominant sector is security and defense, at 56%. As for the underdeveloped and border areas category, the dominant sector is also security and defense, with a weight of 52%.

Analysis of the Role of Perintis Shipping in the 3TP Area

The assessment of the role of Perintis Shipping in the 3TP area is carried out by calculating each assessment indicator in each sector, as discussed in the previous sub-section. Next, the aggregate value results from all districts are grouped into three groups of benefit values. The value of each criterion after normalization can be seen in Table 5. The high value is $x \geq 0.016$; the medium value is $0.011 < \text{value} < 0.016$, and the low value is ≤ 0.011. The assessment of each indicator was carried out for all 3TP areas, so that the following results were obtained in Table 6. The Tanimbar Islands district shows the value of the benefit is 0,0188, which is included in the high benefits category.

Table 5 Value each criterion after normalization

No	District	Category	Political	Economic	Social-cultural	Security and defense	Agregat value
1	Aceh Singkil	Underdeveloped	0.0384697	0.0157698	0.0155318	0.0202020	0.0182
2	Kepulauan Mentawai	Underdeveloped	0.0153879	0.0138942	0.0148439	0.0202020	0.0150
3	Sambas	Underdeveloped	0.0023082	0.0139669	0.0153217	0.0202020	0.0140
4	Rote Ndao	Underdeveloped	0.0139890	0.0172207	0.0155836	0.0202020	0.0166
5	Sabu Raijua	Underdeveloped	0.0128232	0.0179102	0.0160478	0.0202020	0.0169
6	Ende	Underdeveloped	0.0051293	0.0189488	0.0167970	0.0202020	0.0170
7	Manggarai Timur	Underdeveloped	0.0153879	0.0144684	0.0150594	0.0202020	0.0153
8	Sumba Timur	Underdeveloped	0.0063238	0.0174544	0.0158654	0.0202020	0.0161
9	Nabire	Underdeveloped	0.0054957	0.0171545	0.0158401	0.0202020	0.0159
10	Timor Tengah Utara	Underdeveloped	0.0045258	0.0133166	0.0150929	0.0101010	0.0129
-	-	-	-	-	-	-	-
65	Kepulauan Tanimbar	Underdeveloped & Outer/Border	0.0174396	0.0178102	0.0159526	0.0202020	0.0188
		Total	1.0000	1.0000	1.0000	1.0000	

4 Conclusion

Indicators for assessing the role of Pereintis shipping services are seen from several aspects such as politics which is assessed with the connectivity index in 3TP area, economics, which is assessed with correlations between Perintis Shipping voyage and GDRB, social-cultural which is assessed with correlation between Perintis Shipping voyage and HDI, security and defense which is assessed with islands area category. The Tanimbar islands district is included in the category of underdeveloped and outer/border region, which has the percentage of each criterion politics is 26%, economics is 14%, social-cultural is 8%, and security and defense is 52%. From the results of the benefit calculation, the Tanimbar Islands district is a region that has a high benefit value, namely 0.0188.

References

1. C. Amin, H. Mulyati, E. Anggraini, T. Kusumastanto, Impact of maritime logistics on archipelagic economic development in eastern Indonesia. Asian J. Shipp. Log. **37**, 157–164 (2021). https://doi.org/10.1016/j.ajsl.2021.01.004
2. R. Verhaeghe, R.A. Halim, L. Tavasszy, Optimizing the efficiency of the future maritime transport network of Indonesia. Freight Trans. Model. Emerg. Count, pp 109–134 (2021). https://doi.org/10.1016/B978-0-12-821268-4.00006-X
3. T. Achmadi, P. Wuryaningrum, A. Rizki, Evaluation models of effectiveness and efficiency of pioneer ship services: Case study of west papua region. Int. Conf. Marine Technol. (2021)
4. J.S. Park, Y.J. Seo, M.H. Ha, The role of maritime, land, and air transportation in economic growth: Panel evidence from OECD and non-OECD countries. Res. Transp. Econ. **78**, 100765 (2019). https://doi.org/10.1016/j.retrec.2019.100765
5. M. Özer, Ş Canbay, M. Kırca, The impact of container transport on economic growth in Turkey: An ARDL bounds testing approach. Res. Transp. Econ. (2020). https://doi.org/10.1016/j.retrec.2020.101002
6. J. Preston, Integrating transport with socio-economic activity–A research agenda for the new millennium. Trans. Geo. **9**, 13–24 (2001)
7. Ministry of Transportation RI. (2022)

Gap and Discriminant Analysis of the Vessel Service Satisfaction on Coal Ship-To-Ship Services in the Port of Semayang Balikpapan

Anwar Ahmad and Setyo Nugroho

Abstract The Indonesian Statistics Bureau reported a 5.01% (year-on-year) increase in the Indonesian economy in the first quarter of 2022 compared to 2021. The role of ports is one of the factors influencing Indonesia's economic growth. Port activities are critical to the growth of industry and international trade, including both export and import trade, as well as business segments that contribute significantly to Indonesia's development. The Semayang port is a crucial economic gateway in the East Kalimantan province, one of Indonesia's largest coal producers. In this study, a gap analysis was performed to examine the development of coal ship-to-ship (STS) services to improve the infrastructure in distributing new stones throughout Indonesia. To improve existing vessel services at Semayang port, a discriminant analysis of the level of satisfaction with the services provided thus far is required. The result indicated that waiting time and the number of pilots impact ship services the most. The discriminant equation shows that all predictor variables are dominant for predicting differences in customer satisfaction groups with a 57% accuracy. For each survey result, the satisfaction level and expectations are higher than the average, while service users' satisfaction and expectations are lower than the average.

Keywords Gap analysis · Discriminant · Level of satisfaction · Vessel services

A. Ahmad (✉) · S. Nugroho
Marine Transportation Engineering, Sepuluh Nopember Institute of Technology, Surabaya, Indonesia
e-mail: anwarachmad426@gmail.com

S. Nugroho
e-mail: snugroho@seatrans.its.ac.id

A. Ahmad
Shipping and Transport, Rotterdam University of Applied Sciences, Rotterdam, Netherlands

R. Sjarief Widjaja et al. (eds.), *Marine Technology 4*,
SpringerBriefs in Applied Sciences and Technology,
https://doi.org/10.1007/978-3-032-08840-6_13

1 Introduction

Indonesia has 2060 ports; according to the Directorate General of Sea Transportation (2000), there are two categories of ports in Indonesia: public ports, which cater to the needs of the broader public, and special ports, which serve specific needs. Regarding public ports, based on Government Regulation No. 61 of the Republic of Indonesia Year 2009 concerning ports, the hierarchy of port types is divided into three: main ports, collection ports, and feeder ports. The Semayang port is included in the type of public port because it serves inter-island routes and is included in the main port type [1]. Semayang port is a very important port for trade activities and makes trade facilities in the city of Balikpapan increasingly developed, strengthening the function of Balikpapan as a trading city with services covering the entire East Kalimantan region (PT Pelindo IV Balikpapan Branch, 2016). PT Pelabuhan Indonesia Regional four Balikpapan is responsible for developing and managing the Semayang port.

Semayang port strives to provide good and satisfying service to service users to maintain the trust that has been given to the community so far; therefore, as an effort to improve service quality, Semayang port needs to know how far the level of satisfaction of Semayang port service users with the ship services provided so far. One of the efforts to improve the quality of a service is to survey customer satisfaction levels. Questionnaires were used to collect information about services on ships that had to follow certain sizes, like those with a gross tonnage of more than 500 GT [2], as required by the Minister of Transportation Regulation PM 57 of 2015. With this research, it is hoped that Semayang port can plan a better strategy in the future and further improve the quality of its services to meet the wants and needs of service users.

2 Methodology

This chapter explains the theory used in this study.

2.1 Data Sources

In this study, questionnaires or surveys were used to collect data. There were 44 data points collected. Table 1 describes the data scale of each variable on a scale of 1–5. Furthermore, Table 2 describes the data scale of each variable, that is ordinal and ratio-scaled.

Table 1 Variables for closed questions (scale of 1 to 5)

No	Variables
1	Waiting time
2	Number of pillotage
3	Scout skill and expertise
4	Appliction of HSSE and K3 guidance
5	The number of pilot boat
6	The number of Tugboat
7	Visual condition and maintenance of tugboats
8	Main facilities available on tugboats
9	Tugboat performance and service quality
10	Ship service application process
11	Compatibility between rates and services
12	Process of submitting complaints on services

Table 2 Research variables with open questions

Variables	Data scale
Customer satisfaction (Y)	Ordinal
Waiting Time (X_1)	Rasio
The Number of Pilotage (X_2)	Rasio
The number of pilotage boat (X_3)	Rasio
The number of tugboat (X_4)	Rasio

2.2 Gap Analysis

Demonstrates the techniques used for the gap analysis. In Fig. 1, $\overline{Y}$ is the average of the expectations that consumers want while $\overline{X}$ is the average of the perceived reality of consumers.

2.3 Discriminant Analysis

The discipline analysis model is a mathematical equation that represents a linear combination of multiple independent variables [3], as follows:

$$Z = b_0 + b_1 x_1 + b_2 x_2 + \cdots + b_k \quad (1)$$

where Z is the discriminant score and b is the discriminant coefficient or weight, and X is the predictor or independent variable.

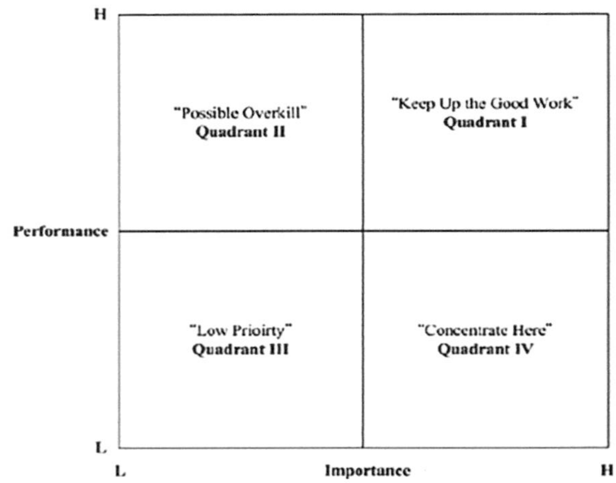

Fig. 1 Cartesian diagram of gap analysis

2.4 *Classification Accuracy*

The calculation of classification accuracy for linear discriminant analysis is based on the apparent error rate (APER), as follows:

$$\text{APER} = \frac{n_{01} + n_{10}}{n} \quad (2)$$

$$\text{Classification Accuracy} = (I - \text{APER}) \times 100\% \quad (3)$$

2.5 *Data Analysis*

The survey data were analyzed using a descriptive approach. Then, the validity and reliability of the survey results were analyzed. Furthermore, GAP analysis was carried out using the important performance analysis (IPA) method. Lastly, discriminant analysis was conducted using Minitab software version 19.

Table 3 Descriptive statistics of research variables

Variables	Mean	Variance	Minimum	Maximum
X_1	33.75	72.24	30	60
X_2	5.00	0.65	4	4
X_3	3.07	0.72	2	2
X_4	12.45	3.23	10	10

3 Results and Discussion

3.1 Descriptive Statistics

The acquired information encompasses the dimensions of data centralization, data distribution, and the inclination of a data cluster [4]. For a more specific illustration, Table 3 Descriptive statistics of research variables provides an elucidation of the extent of data concentration about each research variable. Furthermore, by the results derived from the conducted survey, as illustrated in, it has been established that the clientele's level of contentment with the services offered at Semayang port Balikpapan, surpasses that of dissatisfied patrons by a substantial margin of 64%. Specifically, 64% of respondents expressed satisfaction, while 34% conveyed dissatisfaction.

3.2 Data Quality Test

Firstly, validity tests serve the purpose of ascertaining whether the indicators within a research instrument can accurately measure the intended variables and faithfully capture data [5]. The computed r-count value surpasses the critical table r value of 0.2512. Consequently, it is judicious to conclude that the question indicators employed in the study exhibit validity. Furthermore, the reality test, which constitutes a reliability assessment, endeavors to determine whether the research instrument can be deemed consistent and dependable. This assessment hinges on the stability and consistency of respondents' answers over time.

Notably, the study yielded a Cronbach's alpha score of 0.866. This observed value surpasses the prescribed minimum threshold of 0.6, as stipulated by Cronbach himself. As a result, one may reasonably conclude that the study tool used to assess customer satisfaction variables has a certain degree of reliability.

3.3 Gap Analysis

Provides a visual representation of the positioning of survey questions within different quadrants. Notably, questions Q-1, Q-3, Q-4, and Q-11 are situated in quadrant II, signifying that both satisfaction and expectation values for these inquiries exceed the overall survey average for service user satisfaction and expectations. In contrast, questions 6 and 9 are located in quadrant I, indicating a level of satisfaction below the overall average satisfaction level but above the average expectations. Service providers must address these specific questions to elevate the quality of service rendered to users. Questions 7, 8, 10, and 12 find their place in quadrant III, where satisfaction and expectation scores fall below the mean value for both variables. Finally, in quadrant IV, Q-2, and Q-5 represent regions where satisfaction scores surpass the average value of survey results but remain below the average value of service user expectations.

3.4 Discriminant Analysis Assumption Test

Normal Multivariate Assumption

The hypotheses employed in this study are as follows.

H_0: Multivariate normally distributed data

H_1: Non-multivariate normally distributed data

The evaluation of the normal multivariate assumption is elucidated through the QQ-plot displayed in. Upon careful examination of, it becomes evident that the scatter plots tend to demonstrate linear patterns. This visual observation suggests that the data conforms to a normal distribution. Moreover, additional insights can be gleaned by considering the correlation value in this context. The results of the analysis, where a substantial correlation coefficient of 0.977 is observed, signify a remarkably strong relationship between the variables under consideration. The statistical significance of this correlation is assessed through the p-value, with a threshold typically set at 0.05. In this context, the obtained p-value of 0.00 is substantially lower than the significance level of 0.05, leading to the rejection of the null hypothesis (H_0) or the presumption of multivariate normality in the dataset.

Assumption of Homogeneity

The hypothesis employed in this study is as follows.

$$H_0 : \Sigma_1 = \Sigma_2 = \Sigma_3 = \Sigma_4; H_1 : \Sigma_1 \neq \Sigma_2 \neq \Sigma_3 \neq \Sigma_4 \quad (4)$$

The data obtained from the homogeneity test suggests that a significance level of 5% was utilized. The obtained p-value of 0.694 surpasses the critical value of 0.05, thereby indicating that there is insufficient evidence to reject the null hypothesis (H_0).

Mean Value Vector Assumption

The hypothesis employed in this study is as follows.

$$H_0 : \mu_1 = \mu_2 = \mu_3 = \mu_4; H_1 : \mu_1 \neq \mu_2 \neq \mu_3 \neq \mu_4 \tag{5}$$

According to the mean value vector test, when using a predetermined significance level of 5%, the calculated p-value is 0.750, which is higher than the specified threshold of 0.05. Therefore, the null hypothesis (H_0) remains unchanged, as it cannot be refuted.

Multicollinearity Assumption

As indicated in the multicollinearity test, it is observed that all variables exhibit VIF values below the threshold of 10. This implies that there is an absence of multicollinearity among the variables under consideration, or that the multicollinearity assumption has been satisfied.

Linear Discriminant Analysis

Based on the canonical discriminant function test, it clearly shows that there is a significant difference in the average values of the dependent variable between the two groups for each variable when analyzed separately. Furthermore, according to the canonical discriminant function, the p-values corresponding to variables X_1 through X_4, all surpassing the 0.05 significance level, indicate statistically significant disparities in the means between the two categories for each of these variables. Furthermore, the discriminant function defined for this objective is represented in the following manner:

$$Z = -3.353 + 0.074x_1 + 0.900x_2 - 0.152x_3 - 0.255x_4 \tag{6}$$

Model Goodness

The canonical discriminant function shows that the squared canonical correlation is 0.2172, similar to 0.05. This number shows that the discriminant function can explain

5% of regional origin group diversity. The remaining 95% of variability comes from model-unaccounted sources. Moreover, the discriminant loading defines a variable as having a significant impact on classification if its absolute loading value exceeds 0.40. Thus, parameters X_1 and X_2 significantly and statistically affect categorization.

Model Classification

Based on the canonical discriminant function, it is possible to derive the following outcomes.

$$Z_{CS} = \frac{Z_A + Z_B}{2} = \frac{-0.287 + 0.164}{2} = -0.06 \tag{7}$$

Based on the derived computational outcomes, it is reasonable to deduce that a discriminant score below −0.06 aligns with category 0, designated as "Not Satisfied." Conversely, a discriminant score exceeding −0.06 corresponds to category 1, herein termed the "Satisfied" category.

The values of APER can be determined as follows:

$$\text{APER} = \frac{12 + 7}{9 + 7 + 12 + 16} = \frac{19}{44} = 0.43 = 43\% \tag{8}$$

$$\text{ACCURACY} = 1 - \text{APER} = 1 - 0.43 = 0.57 = 57\% \tag{9}$$

The classification error of the data is 43%. Nevertheless, the level of accuracy attains 57%. The degree of precision is regarded as moderately high, indicating that the discriminant function effectively categorizes customer satisfaction concerning the facilities offered at Semayang port.

4 Conclusion

Based on the analysis undertaken, the following conclusions can be made.

1. This study identifies the primary contributing variables in classification as X_1 (waiting time) and X_2 (the number of pilotage).
2. Based on the discriminant equation, it is evident that all predictor variables exert a significant influence in predicting variations among customer satisfaction groups, yielding an accuracy rate of 57%.
3. The survey results indicate that the satisfaction values and expectations of service users exceed the average levels. In particular, the level of satisfaction is lower than the average satisfaction score and higher than the average expectations score. Hence, service providers should recognize these findings and strive to improve the quality of services offered to service customers.

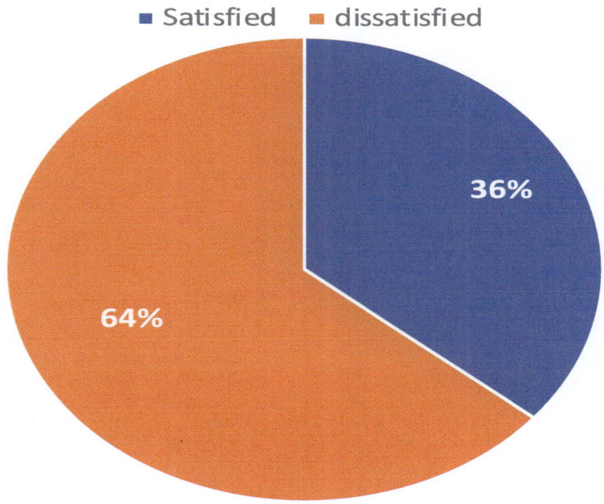

Fig. 2 Customer satisfaction pie chart

The discriminant function has a restriction in that it can only account for 5% of the observed variability. Based on the assumptions mentioned before, it is possible to propose that the implementation of an improved system at Semayang port is anticipated in response to the persisting customer dissatisfaction with the services provided.

1. It is anticipated that future researchers will reassess alternative variables for assessing customer satisfaction that may serve as research variables.
2. According to the gap analysis, the level of satisfaction falls below the average satisfaction while surpassing the average level of expectations. Consequently, service providers must acknowledge this discrepancy to enhance the quality of service provided to service users.
3. It is advisable to ensure that the assumptions underlying discriminant analysis are met to enhance the efficacy of the resulting discriminant model (Figs. 2, 3 and 4).

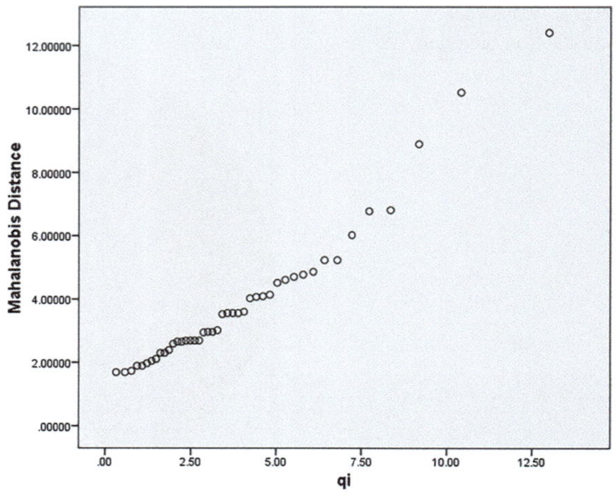

Fig. 3 Cartesian diagrams of importance and performance result

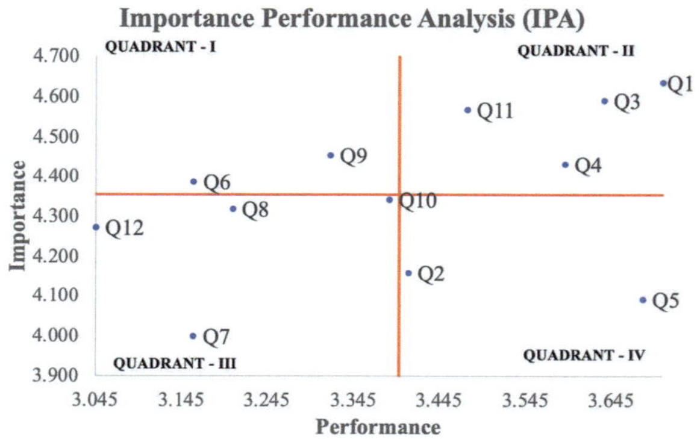

Fig. 4 Normality test graph

Acknowledgements Data for this study were acquired through a survey designed to gauge user satisfaction with ship services at Semayang port, Balikpapan. The survey encompassed a diverse set of stakeholders, including shipping lines/agents, TUKS entities, public ports, and the internal Semayang port.

References

1. B.P. Statistik, Pertumbuhan Ekonomi Indonesia Triwulan I-2022. https://www.bps.go.id/id/pressrelease/2022/05/09/1912/pertumbuhan-ekonomi-indonesia-triwulan-i-2022.html (2022). Accessed 4 June 2022
2. Kementerian Perhubungan Republik Indonesia (Kemenhub RI): Peraturan Menteri Perhubungan Republik Indonesia Nomor 57 Tahun 2015 tentang Pemanduan dan Penundaan Kapal. Database Peraturan JDIH BPK https://peraturan.bpk.go.id/Details/103433/permenhub-no-57-tahun-2015 (2015). Accessed 5 June 2022
3. R.E.Walpole, Pengantar Statistika. PT Gramedia Pustaka, Jakarta (1995)

4. P.K. Faarup, K. Hansen, *Market research and statistics* (Academica, Copenhagen, 2010)
5. N.K. Malhotra, F.M. Ulgado, J. Argawal, G. Shainesh, L. Wu, Dimensions of service quality in developed and developing economics: multi-country cross-cultural comparisons. Int. Mark. Rev. **22**(3), 256–278 (2005)

Logistic Intelligence Boat System Improvement

Wasis Dwi Aryawan, Achmad Zubaydi, Tri Achmadi, and Haikal Anjasmara

Abstract In response to the difficulties of distributing supplies to difficult-to-reach areas, this study proposes an innovative solution by integrating autonomous systems within the vessel to ensure operational safety and efficiency. The previous logistic intelligence boat was developed using a spiral design methodology, leveraging the Bahari-11 as a parent ship for its foundation. The X-bow hull form is also used to improve the wave-making resistance of the ship. The autonomous test employs a waypoint guidance system, validating the vessel's ability to autonomously navigate to a target location. This research focuses on the improvement of the autonomous system of the logistic intelligence boat, showcasing its potential to revolutionize logistics and supply distribution in challenging maritime environments. By integrating excellent autonomous capabilities and leveraging a carefully crafted hull form inspired by the Bahari-11, this innovative vessel offers a robust solution to logistical challenges in remote regions. Overall, this study contributes to the advancement of maritime operations and underscores the significance of autonomous technologies in enhancing efficiency and safety in logistical endeavors.

Keywords Autonomous system · X-bow · Logistic · Waypoint

W. D. Aryawan (✉) · A. Zubaydi · H. Anjasmara
Department of Naval Architecture, Institut Teknologi Sepuluh Nopember, Surabaya, Indonesia
e-mail: wasis@na.its.ac.id

A. Zubaydi
e-mail: zubaydi@na.its.ac.id

T. Achmadi
Department of Marine Transportation, Institut Teknologi Sepuluh Nopember, Surabaya, Indonesia
e-mail: triachmadi@seatrans.its.ac.id

R. Sjarief Widjaja et al. (eds.), *Marine Technology 4*,
SpringerBriefs in Applied Sciences and Technology,
https://doi.org/10.1007/978-3-032-08840-6_14

1 Introduction

The efficient distribution of supplies and resources to remote and challenging maritime environments remains a significant logistical hurdle in today's globalized world. These regions often lack adequate infrastructure and face adverse weather conditions, making traditional supply chain methods unreliable and costly. As a result, there is a growing demand for innovative solutions that can enhance the efficiency and safety of logistics operations in these areas.

In response to these challenges, the integration of autonomous systems within maritime vessels has emerged as a promising and transformative approach. Autonomous vessels have the potential to navigate difficult waters, access remote locations, and optimize supply chain operations. This technology not only offers the promise of greater efficiency but also reduces the risks associated with human-operated vessels in challenging maritime environments.

This research endeavors to address the pressing need for improved logistics solutions in remote and challenging maritime environments. The primary objectives of this study are twofold: Firstly, this research aims to tackle one of the critical impediments to efficient maritime logistics: hull resistance. Traditional vessel designs often encounter high levels of resistance, leading to increased fuel consumption and reduced operational efficiency. To mitigate this issue, we have undertaken a comprehensive hull redesign of the logistic intelligence boat, drawing inspiration from the Ulstein X-bow model, renowned for its ability to reduce resistance [1–3]. Secondly, this study focuses on enhancing the capabilities of the autonomous systems integrated into maritime vessels. Specifically, we explore the development of an advanced autonomous system, featuring a waypoint guidance system that enables the vessel to autonomously navigate predetermined routes and locations. By strengthening the autonomous capabilities of the vessel, we aim to revolutionize supply distribution in challenging maritime environments.

Through these objectives, this research endeavors to contribute to the advancement of maritime operations by demonstrating the potential of autonomous technologies to enhance logistical efficiency and safety. By combining innovative hull design and advanced autonomous systems, we seek to offer a robust solution to the logistical challenges faced in remote maritime regions, thereby paving the way for a more sustainable and effective future in maritime logistics.

2 Methodology

The foundation of this research lies in the development of a prototype autonomous vessel, known as the logistic intelligence boat, which serves as the focal point for our investigation. The initial prototype was created as a comparative study to the reference vessel, Bahari-11, which was previously utilized for the logistic intelligence boat research [4]. This comparison provides valuable insights into the potential improvements achievable through autonomous technology.

A critical component of our methodology involves addressing the issue of hull resistance, a common obstacle to efficient maritime logistics. To overcome this challenge, we initiated a comprehensive hull redesign, drawing inspiration from the Ulstein X-bow model. The X-bow design is renowned for its ability to reduce resistance [1–3], making it an ideal candidate for improving the efficiency of our prototype.

To quantify the impact of our hull redesign, we conducted a rigorous resistance analysis. This analysis involved employing the experimental towing test method, which assesses the vessel's performance under controlled conditions. For testing purposes, a scaled-down model of the vessel was constructed using plywood and fiber materials. This model was then subjected to the towing test, allowing us to measure the effectiveness of the hull redesign in reducing resistance. The results of this analysis provide critical data to evaluate the performance improvements achieved through our design modifications.

The second major component of our methodology centers around enhancing the autonomous systems integrated into the logistics intelligence boat. These systems are pivotal in enabling the vessel to operate independently and navigate challenging maritime environments effectively.

A key innovation in autonomous systems is the implementation of a waypoint guidance [5] and collision avoidance system [6, 7]. This system allows the vessel to autonomously follow predefined routes and navigate to specific waypoints while actively avoiding collisions with obstacles. It is designed to optimize the vessel's path, taking into account factors such as weather conditions, obstacles, and fuel efficiency. The waypoint guidance and collision avoidance system holds the potential to significantly enhance the vessel's ability to autonomously operate in remote and challenging maritime environments, ensuring both efficient navigation and the safety of the vessel and its surroundings. By integrating the waypoint guidance system, we aim to demonstrate the practical applicability and benefits of autonomous navigation in the context of logistics and supply distribution.

3 Results and Discussion

3.1 Hull Redesign Outcomes

The hull redesign based on the X-bow model resulted in a significant reduction in resistance, ultimately enhancing fuel efficiency. The redesigned hull model can be seen in Fig. 1, and the differences between the previous design and the new design can be seen in Fig. 2.

Fig. 1 Redesign hull model

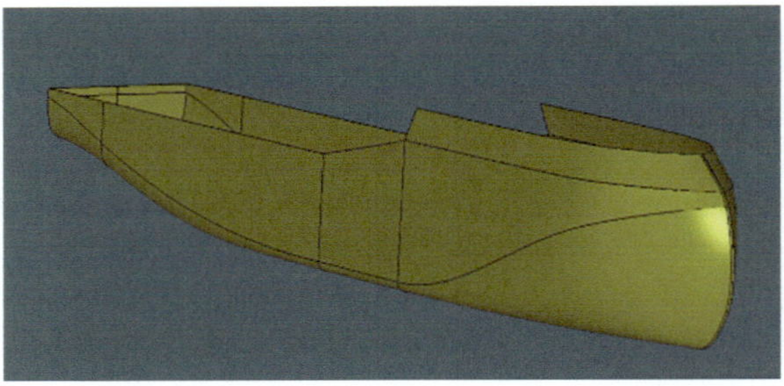

(a) Previous Model [4]

(b) X-bow Model

Fig. 2 Redesign outcomes **a** previous model, **b** X-bow model

3.2 Resistance Analysis

The experimental towing test revealed the effectiveness of the hull redesign in reducing resistance, confirming the benefits of the X-bow-inspired design. The model was tested under calm water conditions, where the final estimate of propulsive power is in the form of EHP (effective horsepower). To determine the size of the engine (BHP), the measurement results must be corrected by including corrections for the efficiency of the hull, propeller shaft, propeller, rudder, the type of water in which the ship will operate, and other factors.

Variations in the relationship between ship speed (knots) and total ship resistance (kN) are illustrated in Fig. 3. There is an increase in resistance as the ship's speed increases. Variations in the relationship between speed (knots) and the magnitude of ship propulsive power (EHP) are displayed in Fig. 3 and Table 1, where an increase in the ship's EHP is observed as the ship's speed increases. This is a direct result of the increase in total ship resistance associated with the increase in ship speed.

An increase in resistance force tends to occur linearly from the initial or lowest speed of the model, which is 0.7413 m/s (corresponding to a ship speed of 9 knots), to the final or highest speed of the model, which is 1.1532 m/s (corresponding to a ship speed of 14 knots). From this information, it is possible to estimate the ideal service speed of the ship and the magnitude of the ship's EHP.

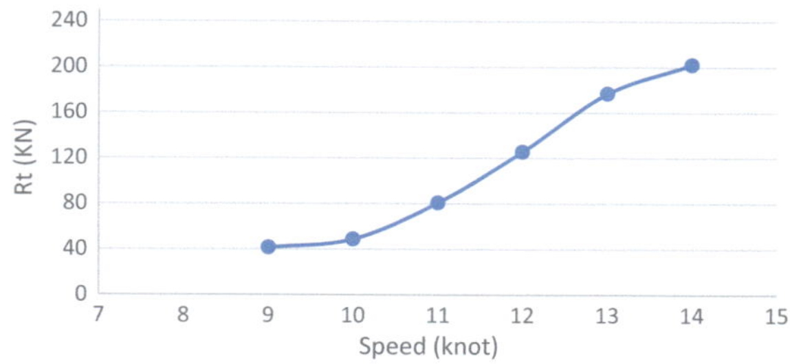

Fig. 3 Resistance analysis

Table 1 The relationship between service speed, resistance (RT), and ship's fffective horsepower (EHP)

Run	V (knot)	Rt (KN)	Power(EHP)	Cf	Ct	Rn
1	9.0000	41.5795	261.7952	0.001809	0.037955	274,352,085.97
2	10.0000	48.8704	341.8895	0.001784	0.036134	304,835,651.07
3	11.0000	80.7835	621.6634	0.001761	0.049364	335,319,216.18
4	12.0000	125.6396	1054.7457	0.001741	0.064511	365,802,781.29
5	13.0000	176.9507	1609.2940	0.001723	0.077417	396,286,346.40
6	14.0000	202.2724	1981.0918	0.001706	0.076305	426,769,911.50

3.3 Autonomous System

The upgraded autonomous system demonstrated its capability to autonomously navigate to predetermined waypoints, showcasing its potential to improve logistics and supply distribution in challenging maritime environments. An autonomous system can be divided into two main components: power architecture and communication architecture. The autonomous system block diagram of communication and power architecture can be seen in Figs. 4 and 5.

The autonomous system's power architecture manages power distribution, ensuring components receive adequate energy for operation. It includes designing and distributing power sources like batteries or solar panels. Communication architecture orchestrates interactions within the system, defining data exchange pathways and protocols. This facilitates effective command transmission, feedback reception, and coordination among components, crucial for real-time decision-making.

In summary, within an autonomous system, power architecture deals with the management of power sources and distribution, while communication architecture handles the flow of control and data between the main control system and other components. These two components work together to ensure the system operates autonomously and efficiently.

The components used in the autonomous system block diagram are listed and can be seen in Table 2. The autonomous system's block diagram depicts various critical components required for its smooth operation. The single-board computer (SBC) serves as the system's brain, receiving sensor input, processing it using algorithms,

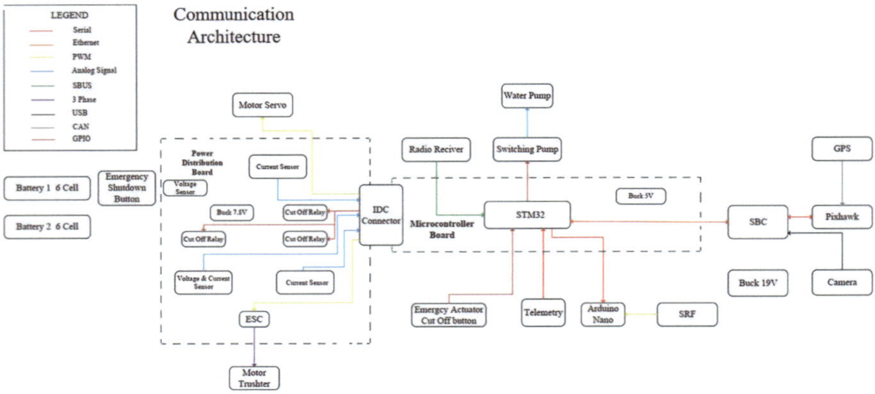

Fig. 4 Autonomous system block diagram—communication architecture

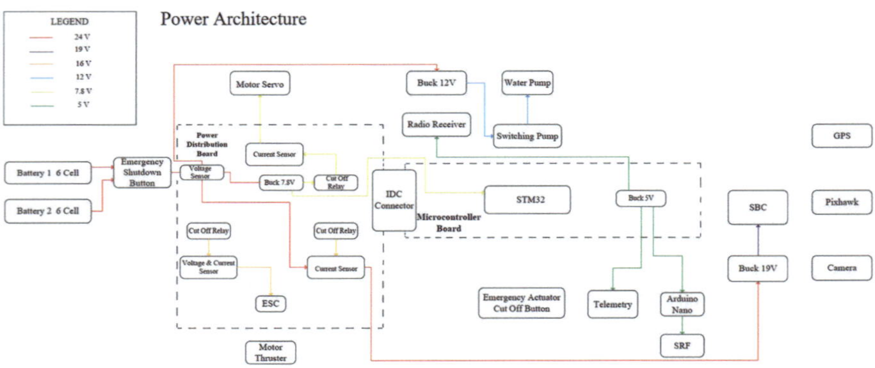

Fig. 5 Autonomous system block diagram—power architecture

and making informed decisions. The STM32 microcontroller serves as the core processing unit, utilizing its computational capabilities to handle real-time control and decision-making activities. Meanwhile, the Pixhawk autopilot serves as the core processing unit, directing the autonomous vehicle's navigation and control, as well as supporting multiple flight modes and propulsion system management. GPS receivers serve an important role in establishing the system's position and velocity, which is required for navigation and localization. Camera-based obstacle avoidance increases safety by identifying and avoiding impediments in real time. Other key components include motor servos for precise control, motor thrusters for underwater propulsion, and sensors such as speed encoders and steering sensors to measure movement and steering parameters. Finally, the buck converter ensures effective power management by modulating voltage to fit the unique needs of each component. These components work together to build a strong basis for autonomous operation, allowing for efficient and reliable operations, including navigation, obstacle avoidance, and precise control.

Table 2 Autonomous system components

No	Item
1	Single board computer (SBC)
2	Microcontroller
3	Pixhawk
4	GPS
5	Camera
6	Motor servo
7	Motor thruster
8	Speed encoder
9	Steering sensor
10	Battery
11	Buck converter

In summary, these components collectively enable the autonomous system to function effectively. They process data, make decisions, provide precise positioning, avoid obstacles, control movement, and manage power distribution, all working together seamlessly to support autonomous operation.

4 Conclusion

The logistics intelligence boat has undergone a significant redesign, featuring a rebuilt autonomous system. Key insights from this redesign and development process include:

1. Utilizing the X-bow hull form effectively reduces the ship's resistance.
2. The autonomous system incorporates essential components such as control units, sensors, actuators, and power supplies. It also integrates ship-specific elements like servo motors, thruster motors, speed encoders, and buck converters to enhance its capabilities.

Acknowledgements The author would like to thank the Ministry of Education and Culture which funded this research.

References

1. A.P. Sidabalok, D. Chrismianto, M. Iqbal, Analisa Perbandingan Ulstein X-Bow Dengan Bulbous Bow Konvensional Terhadap Nilai Hambatan Total Dan SEAKEEPING KAPAL MENGGUNAKAN METODE CFD. J. Tek. Perkapalan. **4**, 141–151 (2016)

2. R. Kuasa, S. Hadi, M. Iqbal, Optimalisasi Curve Linesplan Haluan Kapal Perintis 750 DWT Menggunakan Response Surface Methode (RSM) untuk Mengurangi Hambatan. J. Tek. Perkapalan. **5**, 810 (2017)
3. M, Fatima, K. Deddy Chrismianto, Analisis Pengaruh Variasi Sudut Flare Bow Dengan Stem Terhadap Nilai Hambatan Dan Slamming Pada Model Haluan Ulstein X-Bow. J. Tek. Perkapalan Univ. Diponegoro. **6**, 20–27 (2018)
4. W.D. Aryawan, A. Zubaydi, T. Achmadi, H. Anjasmara, N.P.L. Sukma, A. Asmara, A.E. Indaryo, M.I. Fadhok, H. Yusuf, Design of a logistic intelligence boat. IOP Conf. Ser. Earth Environ. Sci. **1198** (2023). https://doi.org/10.1088/1755-1315/1198/1/012032
5. Y.A. Ahmed, K. Hasegawa, Fuzzy reasoned waypoint controller for automatic ship guidance. IFAC-PapersOnLine. **49**, 604–609 (2016). https://doi.org/10.1016/j.ifacol.2016.10.501
6. X. Zhang, C. Wang, L. Jiang, L. An, R. Yang, Collision-avoidance navigation systems for maritime autonomous surface ships: A state of the art survey. Ocean Eng. **235**, 109380 (2021). https://doi.org/10.1016/j.oceaneng.2021.109380
7. Z. Zhou, Y. Zhang, S. Wang, A coordination system between decision making and controlling for autonomous collision avoidance of large intelligent ships. J. Mar. Sci. Eng. **9** (2021). https://doi.org/10.3390/jmse9111202

Programming Strategies in AFM Lithography: Effects on Mask Oxidation in Silicon Nanowire Transistor Fabrication

Ahmad Makarimi Abdullah, Khatijah Aisha Yaacob, Mohd Azraei Mohd Azmi, and Asrulnizam Abd Manaf

Abstract Atomic force microscopy (AFM) lithography is a highly precise nanofabrication technique essential for silicon nanowire transistor (SiNWT) development. The programming strategies used in AFM lithography directly impact mask oxidation, influencing the overall performance and reliability of SiNWTs. This study explores how different AFM programming sequences affect oxidation by optimizing key parameters such as tip bias voltage, scan speed, and environmental conditions. The findings demonstrate that well-controlled programming strategies improve oxidation uniformity and pattern fidelity, leading to enhanced SiNWT functionality. Additionally, this study provides insights into oxidation kinetics influenced by AFM parameters, contributing to better control over nanoscale patterning. By refining AFM lithography techniques, semiconductor device fabrication can achieve higher precision, reproducibility, and efficiency. These results highlight the repeatability and reproducibility of the SiNWT device fabrication to enhance the potential of AFM lithography in advancing next-generation nanoelectronic devices in fabrication processes.

A. M. Abdullah · K. A. Yaacob (✉)
School of Material and Mineral Resources Engineering, Engineering Campus, Universiti Sains Malaysia, Nibong Tebal, Penang, Malaysia
e-mail: mraisha@usm.my

A. M. Abdullah
e-mail: makarimi@unikl.edu.my

A. M. Abdullah
Malaysia Institute of Marine Engineering Technology, Universiti Kuala Lumpur, Lumut, Perak, Malaysia

M. A. M. Azmi
British-Malaysia Institute, Universiti Kuala Lumpur, Gombak, Selangor, Malaysia
e-mail: mazraie@unikl.edu.my

A. A. Manaf
Electrical and Electronics Engineering, Engineering Campus, Universiti Sains Malaysia, Nibong Tebal, Penang, Malaysia
e-mail: ; eeasrulnizam@usm.my

R. Sjarief Widjaja et al. (eds.), *Marine Technology 4*,
SpringerBriefs in Applied Sciences and Technology,
https://doi.org/10.1007/978-3-032-08840-6_15

Keywords AFM lithography · Silicon nanowire transistor (SiNWT) · RASTER programming

1 Introduction

The rapid advancement of nanotechnology has enabled the creation of highly efficient and miniaturized semiconductor devices. Among these, silicon nanowire transistors (SiNWTs) have emerged as promising candidates for next-generation electronics due to their excellent electrical properties—such as high carrier mobility and low power consumption—making them ideal for use in high-performance computing, sensing, and logic circuits [1–6]. However, fabricating SiNWTs requires precise nanoscale patterning to ensure structural accuracy and reproducibility [7–9]. Atomic force microscopy (AFM) lithography stands out among available techniques for its capability to directly pattern nanostructures with high precision and control [9, 10]. A major challenge in this method is managing the mask oxidation process, which directly influences the fidelity of the final structure [7, 10]. Oxidation behavior is closely tied to programming strategies such as tip bias voltage, scan speed, humidity, and temperature during AFM operation [11, 12]. This study investigates how different AFM programming sequences affect oxidation during SiNWT fabrication. By analyzing the interplay between lithography parameters and oxidation kinetics, the research aims to optimize AFM nanofabrication. The findings support improved precision and control in nanoscale patterning, ultimately enhancing device performance and reliability in future nanoelectronic applications.

2 Methodology

This study examines the impact of AFM lithography programming sequences on mask oxidation during the fabrication of silicon nanowire transistors (SiNWTs). The methodology consists of four main stages: substrate preparation, AFM lithography programming, oxidation process, and characterization of the oxide patterns, as the process of study is illustrated in Fig. 1.

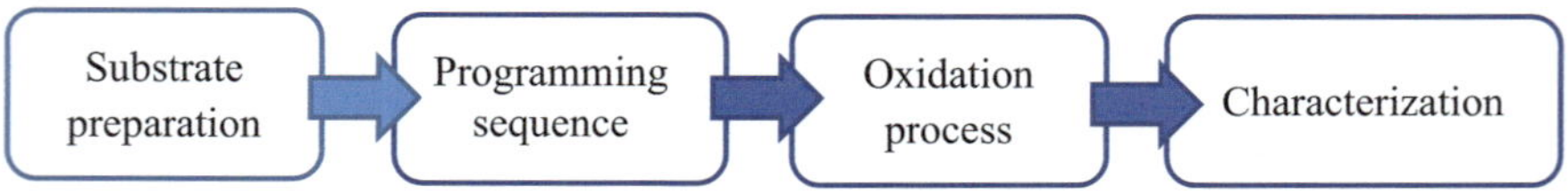

Fig. 1 The methodology of device fabrication

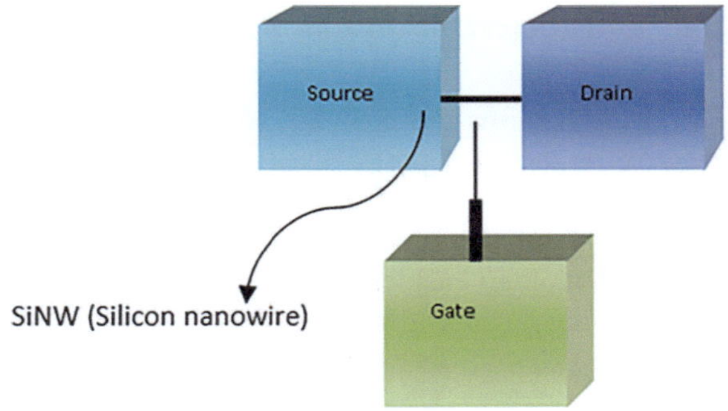

Fig. 2 Illustration of silicon nanowire transistor (SiNWT) design on SOI

2.1 Substrate Preparation

A silicon-on-insulator (SOI) wafer with a thin silicon layer is used as the base substrate. A p-type SOI wafer (510 mΩ·cm, 70 nm Si, 2000 nm SiO_2) was prepared and cut into 1.5 cm^2 samples. Standard RCA1 and RCA2 cleaning removed contaminants, followed by HF (1:80 in DIW) to strip native oxides.

2.2 AFM Lithography Programming (Programming Sequences)

Samples were then placed in an AFM chamber (SPA300HV) for nanolithography. AFM parameters were configured for transistor shaping on SOI. The circuit layout was accurately shaped using raster-based vector scanning commands, as illustrated in Fig. 2. Different programming sequences are applied by varying key parameters such as, (i) *Tip Bias Voltage:* Adjusted between 5 and 20 V to control the oxidation rate, (ii) *Scan Speed:* Varied from 0.1 μm/s to 1 μm/s to examine its impact on oxidation uniformity, (iii) *Environmental Conditions:* Relative humidity in the laboratory is controlled to study its influence on oxide growth.

2.3 Oxidation Process

The oxidation reaction is induced by applying a bias voltage to the AFM tip, forming a localized oxide layer on the SOI substrate. The oxidation process is monitored in real time to assess the effects of different programming strategies. During this stage, AFM lithography is performed using a conductive AFM probe operated in contact mode.

2.4 Characterization and Analysis

The oxide patterns are analyzed using: (i) a*tomic force microscopy (AFM):* To measure oxidation depth and pattern fidelity, (ii) *scanning electron microscopy (SEM):* To observe the surface morphology of the oxidized mask, and (iii) *energy dispersive X-ray spectroscopy (EDS):* To confirm the presence of oxide formation. This methodology enables a comprehensive evaluation of how AFM lithography programming sequences influence mask oxidation, providing insights for optimizing nanoscale patterning techniques for SiNWT fabrication.

3 Results and Discussion

3.1 Influence of Tip Bias Voltage on Oxidation Pattern

The results show a clear link between tip bias voltage and oxidation depth. At lower voltages (5–10 V), oxidation is slower, forming thinner and moderately uniform oxide layers. As the voltage rises to 15–20 V or more, oxidation depth increases, though high voltages can cause irregular patterns due to breakdown effects, affecting lithography precision [13]. This suggests there's an optimal voltage range for consistent oxidation [14, 15]. In this study, tests used a 10 V Cr/Pt tip under stable conditions (22–26 °C, 65–68% RH), yielding reproducible results [14–16]. Figure 3a and b show nanowire widths and heights at various voltages, with dimensions increasing from 171.74 nm at 5 V to 257.61 nm at 20 V, measured at a scan speed of 0.5 μs.

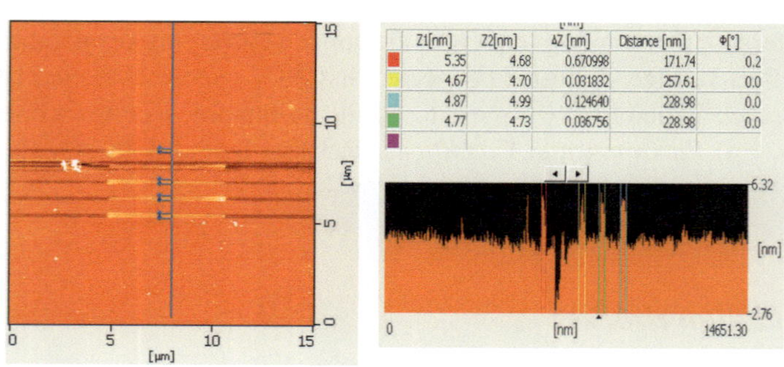

(a) (b)

Fig. 3 AFM image of **a** oxidation patterned (nanowire) with different tip bias voltage (5–20 V), **b** cross-section of nanowire with different applied tip bias voltage

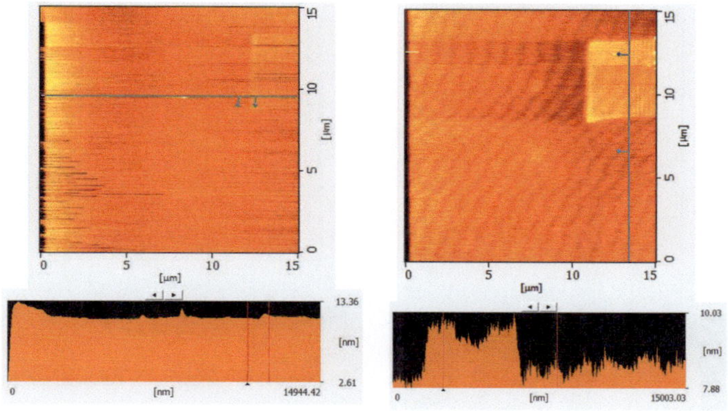

(a) (b)

Fig. 4 AFM image of **a** oxidation patterned (nanowire) with different tip bias voltage (5–20 V), **b** cross-section of nanowire with different applied tip bias voltage

3.2 Effect of Scan Speed on Oxidation Uniformity

The AFM probe's scan speed greatly affects the uniformity of the oxidation pattern. At higher speeds (0.8–1.0 μm/s), oxidation becomes uneven, causing breaks in the oxide layer [17, 18]. Slower speeds (0.1–0.3 μm/s) provide better control but increase processing time. An optimal speed of 0.5 μm/s was found to offer a good balance between uniformity and efficiency, and was used in all sample fabrications. Figure 4a shows that at 1.0 μm/s, the oxide layer is only ~ 0.48 nm—too thin for reliable patterning. In contrast, Fig. 4b shows that at 0.3 μm/s, a more uniform and thicker oxide layer (~1.02 nm) forms, though with a higher risk of excess buildup. All tests were done at 22–26 °C, 65–68% RH, using a 10 V tip bias.

3.3 Role of Environmental Conditions in Oxidation Kinetics

Controlled humidity is vital in the AFM oxidation process. When relative humidity drops below 65%, oxidation slows down, resulting in incomplete patterns. On the other hand, humidity above 70% speeds up oxidation excessively, leading to uneven oxide layers and possible defects. An ideal range of 65%–68% ensures stable, high-quality pattern formation on the SiO_2 mask on the SOI substrate [14, 19]. As shown in Fig. 5a and b, oxide thicknesses of ~ 1.53 nm and ~ 2.47 nm clearly reflect the humidity's effect, under identical conditions.

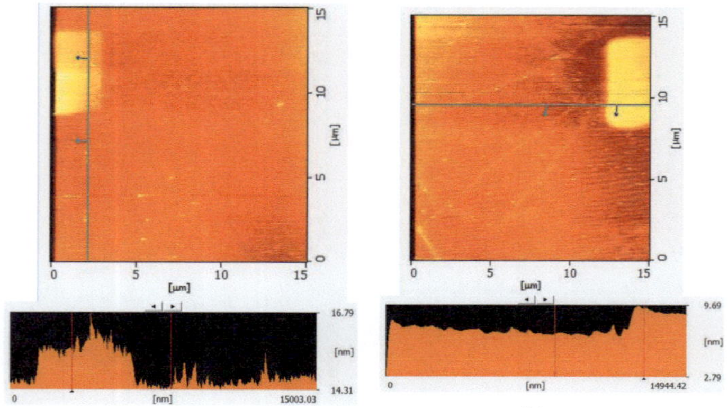

(a) (b)

Fig. 5 Shows AFM images of oxide patterns. In **a**, the oxide thickness is ~ 1.53 nm at 65–68% RH, 0.5 µm/s scan speed, and 10 V bias. In **b**, the thickness increases to ~ 2.53 nm at 70% RH under the same scan speed and bias

3.4 *The RASTER Programming Patterning Effects*

In this AFM oxidation lithography study, the RASTER patterning sequence was used to guide the tip's path and timing. This sequence is crucial for accurately shaping nanostructures, ensuring proper alignment and spacing, especially in complex or large-scale patterns. Poorly optimized programming can lead to uneven exposure, causing inconsistent oxide thickness and blurred pattern edges. When the tip pauses too long in one spot, excess oxidation creates thicker oxides and shadow effects due to lateral diffusion. On the other hand, continuous tip movement without enough pauses may cause oxide lines to merge, reducing resolution. These issues are strongly influenced by oxidation parameters. High tip bias voltage accelerates oxidation but risks overgrowth, while high humidity, though necessary for forming the water meniscus, can increase diffusion effects, especially with prolonged exposure. Improper AFM tip discharge may also leave residual charges, causing unintentional oxidation. Achieving precise results requires a careful balance of bias voltage, scan speed, environmental conditions, and RASTER programming. Figure 6a and b show oxidation shadows caused by the all-at-once approach. In contrast, Fig. 7a and b show improved structures using a step-by-step method (Fig. 8), completed over 90 min.

3.5 *Structural and Morphological Analysis*

AFM imaging confirms that optimized RASTER programming yields sharp, well-defined oxide lines with minimal roughness. SEM analysis shows uniform pattern dimensions, matching the AFM design. EDS results in Fig. 9 validate the oxidation

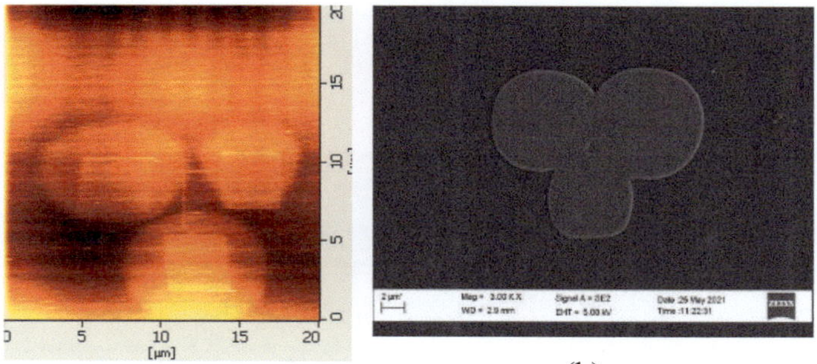

(a) (b)

Fig. 6 All-at-one approach programming for oxidation process **a** AFM image and **b** SEM micrograph picture

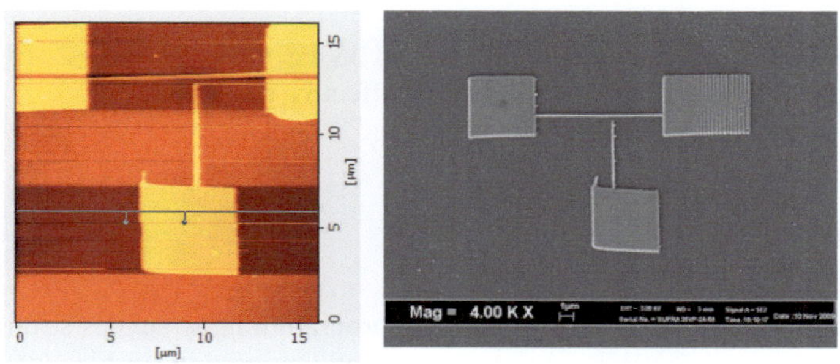

(a) (b)

Fig. 7 Step-by-step approach programming for oxidation process **a** AFM image, **b** SEM micrograph picture

Fig. 8 Step-by-step programming strategies for oxidation process using RASTER programming in AFM oxidation

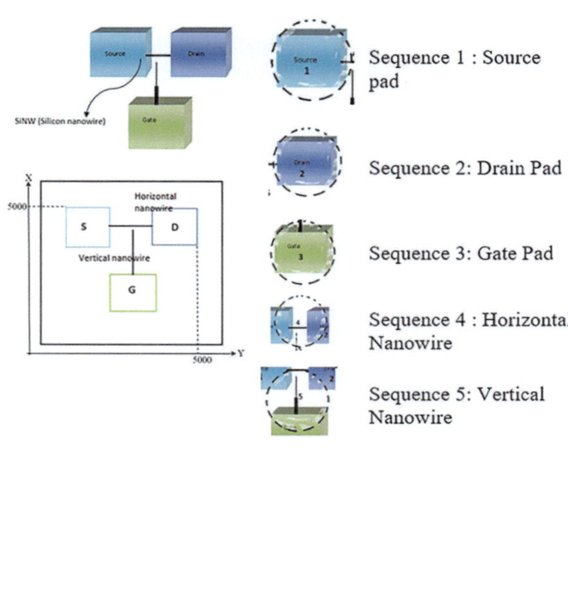

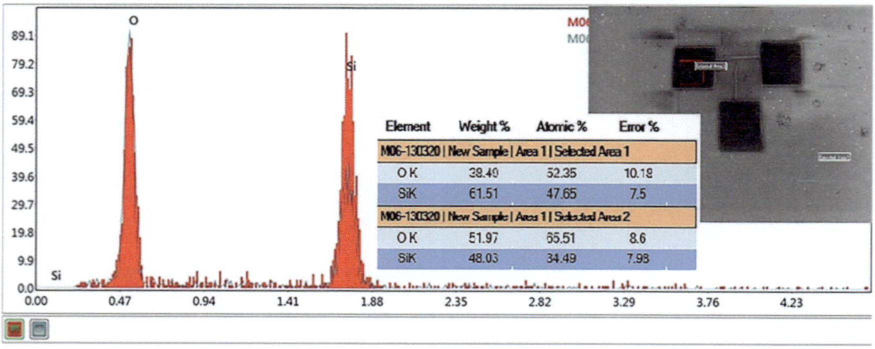

Fig. 9 SEM micrograph picture with EDX analysis for final structure of SiNWT after etched process

process: Sect. 1 (pad) contains 61.5% Si and 38.4% O, while Sect. 2 (BOX layer) shows 48.03% Si and 51.7% O, confirming effective oxidation [14].

4 Conclusion

The findings confirm that AFM lithography programming strategies strongly affect mask oxidation and SiNWT fabrication quality. Optimizing tip bias (10 V), scan speed (0.5 μm/s), humidity (65–68%), and using a step-by-step RASTER approach ensures uniform, precise patterns [7, 10–12]. Structural analysis supports these results, highlighting AFM lithography's reliability for advanced nanoelectronic fabrication.

Acknowledgements The Universiti Sains Malaysia funded this research through a research university grant RU-8014205 under the supervision of A.P. Dr. Khatijah Aisha Yaacob. The Universiti Kuala Lumpur funded for writer study program sponsorship.

References

1. Y. Hashim, W.A. Jabbar, Optimization of n-MOS 6T nanowire SRAM bit cell based on nanowires ratio of SiNWTs. J. Nano-Electron. Phys. **12**(5), 05020 (2020)
2. A.M. Abdullah et al., The effect of nanowire gap for silicon nanowire transistor to the current-voltage (I-Vds). J. Sci. Technol. **15**(2), 1–10 (2023)
3. H.T.Al Ariqi et al., Characterization of silicon nanowire transistor. Telkomnika. **23**(4) (2019)
4. S. Kim, D. Park, Flexible electronic platforms integrated with silicon nanowire transistors. Adv. Mater. Interfaces. **10**(4), 2100885 (2023)
5. A. Mahmood et al., Electrical characterization of Ge-FinFET transistor based on nanoscale channel dimensions. J. Nano Electron. Phys. **11**(1), 1011 (2019)

6. J.I.A. Rashid et al., The development of silicon nanowire as sensing material. Hindawi Publ. **2013**, 14 (2013)
7. M. Wang et al., Self-aligned patterning technique for fabricating high performance diamond sensor arrays with nanoscale precision. Sci. Adv. **8**, 38 (2022)
8. Y. Bae et al., Atomic force microscope-guided nanoscale 3D patterning for carbon nanofibers with in situ Raman spectroscopy. Nanoscale (2025)
9. M. Kim et al., Cracking-assisted fabrication of nanoscale patterns for micro/nanotechnological applications. Nanoscale. **8**(18) (2016)
10. N. Najihah et al., Fabrication of SiNWs-FET nanostructure via atomic force microscopy lithography. Solid State Phenom. **301**, 103–110 (2020)
11. M. Arai et al., Fabrication of 10-nm-scale nanoconstrictions in graphene using AFM-based local anodic oxidation lithography, extended abstracts of the 2014 international conference on solid state devices and materials (2014)
12. H.Y. Zhang, Nano structures constructed by AFM based lithography. Adv. Mater. Res. **815**, 490–495 (2013)
13. S. Bae, C. Han et al., Atomic force microscope anodization lithography using pulsed bias voltage synchronized with resonance frequency of cantilever. Nanotechnology **16**(10), 2082–2085 (2005)
14. A.M. Abdullah et al., Influence of room humidity on the formation of nanoscale silicon oxide patterned by AFM lithography. Int. J. Nanosci. **09**(04), 251–255 (2010)
15. A.M. Abdullah et al., The effect of nanowire gap for silicon nanowire transistor to the current-voltage (I-Vds). J. Sci. Technol. **15**(2) (2023)
16. T. Chen et al., Formation of Ni-silicide nanowires on silicon-on-insulator substrates by atomic force microscope lithography and solid phase reaction. ECS J. Solid State Sci. Technol. **1**(2), P90–P93 (2012)
17. K. Zhang et al., Compensation method for correcting the topography convolution of the 3d AFM profile image of a diffraction grating. Machines **12**(2), 126 (2024)
18. S. Jian, J. Juang, Scanned probe oxidation onp-gaas(100) surface with an atomic force microscopy. Nanoscale Res. Lett. **3**(7) (2008)
19. A. Dehzangi, Larki et al., Atomic force microscope base nanolithography for reproducible micro and nanofabrication. in *2014 IEEE international conference on semiconductor electronics (ICSE2014)*, pp 408–411 (2014)

MIX
Papier aus verantwortungsvollen Quellen
Paper from responsible sources
FSC® C105338

If you have any concerns about our products,
you can contact us on
ProductSafety@springernature.com

In case Publisher is established outside the EU,
the EU authorized representative is:
Springer Nature Customer Service Center GmbH
Europaplatz 3, 69115 Heidelberg, Germany

Printed by Libri Plureos GmbH
in Hamburg, Germany